The Grand Communications Route from Saint John to Quebec

W.E. (GARY) CAMPBELL

GOOSE LANE EDITIONS and
THE NEW BRUNSWICK MILITARY HERITAGE PROJECT

Edited by Brent Wilson.
Cover illustrations: Front: *Reinforcements for Canada Passing Through New Brunswick: A Portion of the 63rd Regiment Crossing Nerepis Valley*, ILN UNB. Back: *Officers' Barracks at Fredericton, Winter, 1834* (1836), by W.P. Kay, NAC C-041071.
Cover and interior design by Julie Scriver.
Printed in Canada by Marquis Book Printing.
10 9 8 7 6 5 4 3 2 1

Library and Archives Canada Cataloguing in Publication

Campbell, W. E. (William Edgar), 1947-
The road to Canada: the grand communications route from Saint John to Quebec / W.E. (Gary) Campbell.
(New Brunswick military heritage series; 5)
Includes bibliographical references and index.
ISBN 0-86492-426-7
1. Roads — New Brunswick — History. 2. Indian trails — New Brunswick — History.
3. Military roads — Canada, Eastern — History.
4. Trans-Canada Highway — History. I. Title. II. Series.
FC2461.R63 2005 971.5'1 C2005-902272-8

Published with the financial support of the Canada Council for the Arts, the Government of Canada through the Book Publishing Industry Development Program, and the New Brunswick Culture and Sports Secretariat.

Goose Lane Editions
469 King Street
Fredericton, New Brunswick
CANADA E3B 1E5
www.gooselane.com

New Brunswick Military Heritage Project
Military and Strategic Studies Program
Department of History, University of New Brunswick
PO Box 4400
Fredericton, New Brunswick
CANADA E3C 1M4
www.unb.ca/nbmhp

Dedicated to my wife
Carolyn
without whose loving support this book would not have been possible
and
to the memory of
my great-great-great-great-grandfather,
Private William Moran, 104th Regiment of Foot
(late the King's Orange Rangers and the King's
New Brunswick Regiment),
who died at Kingston, Ontario, on January 12, 1814,
while on active service,
having made the winter march to Canada,
and two of his sons,
Private John Moran, 104th Regiment of Foot,
who also made the winter march to Canada,
and
Boy and Private William Moran, 104th Regiment of Foot

Contents

The Road to Canada

The Grand Communications Route from Saint John to Quebec

Arrival of a detachment of the 63rd Regiment at the temporary barracks at Petersville, the first stop on the Great Communications Route and now part of Canadian Forces Base Gagetown. ILN UNB

Introduction

"It has been said, with truth, that the history of human civilization has been determined and controlled by great rivers."
— *William O. Raymond,* The River St. John

At noon on Monday, April 12, 1813, a long column of infantry stopped to wash the mud off their legs before marching smartly into the garrison at Kingston, Upper Canada. The 104th Regiment of Foot had left Fredericton, New Brunswick, fifty-two days and over 1,128 kilometres earlier. The march through the wilds of northern New Brunswick in the dead of winter and then along the banks of the St. Lawrence River to Kingston was accomplished without loss of life. One soldier wrote that the men had "marched on snowshoes in one of the severest winters ever known . . . undergoing hardships unequalled by any regiment in service during the war." After they passed Quebec, the weather broke; then they "marched a part of the distance through mud, water and slush, knee deep." They had come to help save the Canadas from invasion. The march of the 104th became the best-known movement of troops along the Grand Communications Route through New Brunswick, but it was only one of many; until the late nineteenth century, this route formed the backbone of the French and then the British empires in North America.

Today, many people are familiar with the importance of North American rivers such as the St. Lawrence and the Mississippi, but few realize that until quite recently the St. John River was also an important route joining the Atlantic to the hinterland of the Continent. This route ran from the mouth of the St. John River on the Bay of Fundy to the junction with the Madawaska River at Edmundston, up the Madawaska northwards to Lake Temiscouata, and on over the Grand Portage to the St. Lawrence River at Rivière-du-Loup; the British came to call this the Grand Communications Route. Until the advent of railways and steamships, from November to April, when ice closed the St. Lawrence River, it was the only secure way for the French and then the British to reach Canada.

The St. John River system was used for personal travel and trade and by hunting and war parties long before the arrival of the Europeans in the seventeenth century. When French kings began exercising control over Acadia through their governor in New France, the route gained strategic significance in the European struggle for empire. While the preferred method of transportation was by ship, sea communication between Acadia and New France was not possible during the winter months. The only way to pass messages or dispatches or to move troops between the two places during the winter was over the Grand Communications Route, and it was used regularly during the rest of the year by couriers and small parties. The French also made extensive use of the branch route from the St. John River through Lake Washademoak, along the Canaan River, and over the Petitcodiac River portage to communicate with Fort Beauséjour, Louisbourg and Port-Royal.

Following the conquest of Acadia and New France in the mid-1700s, the British inherited the route, but for a time they could also use the easier path from Montreal via the Richelieu River, Lake Champlain and the Hudson River to the ice-free port of New York. The American Revolution, however, thrust the Grand Communications Route into prominence once again, and the British made a concerted effort, not just to control it, but to improve it to accommodate sleighs and wagons. The vague definition of the border between Maine and New Brunswick in the 1783 Treaty of Paris made this very difficult. The American interpretation of the boundary placed parts of the Grand Communications Route inside their territory, threatening its integrity, and therefore the British settled Loyalist regiments along its path to help provide security. The route proved its value during the War of 1812, when troops urgently needed to defend

Canada, including the 104th, moved over it during the winters of 1813 and 1814.

After the War of 1812, disbanded British regiments settled along the upper St. John River to protect the route from American incursions. The route proved its importance again during the winters of 1837-1838 and 1838-1839, when it was used to move much needed reinforcements to Canada following the outbreak of rebellion. The Maine-New Brunswick border controversy reached its highest period of tension during the Aroostook War of 1839. Fortunately, open conflict was avoided, and the 1842 Webster-Ashburton Treaty placed the whole route firmly in British territory. Indeed, the British sacrificed the lumber and agricultural riches of Aroostook County, Maine, in order to retain the Grand Communications Route. This strategic goal was validated almost twenty years later during the American Civil War, when the Trent Affair of November, 1861 brought Great Britain and the United States to the brink of war. In the winter of 1861-1862, to deter the American invasion expected in the spring, the British sent the largest-ever reinforcement of troops to Canada along the Grand Communications Route.

Following the peaceful resolution of the Trent Affair and the signing of the Treaty of Washington in 1871, the military importance of the Grand Communications Route faded. However, the route itself remained in use. Roads and railways followed its path from Saint John to the St. Lawrence River and continued the link between the Maritimes and Central Canada. Today, anyone who drives along the Trans-Canada Highway between Rivière-du-Loup and Fredericton is following the Grand Communications Route, which played such a central part in the colonial history of Canada.

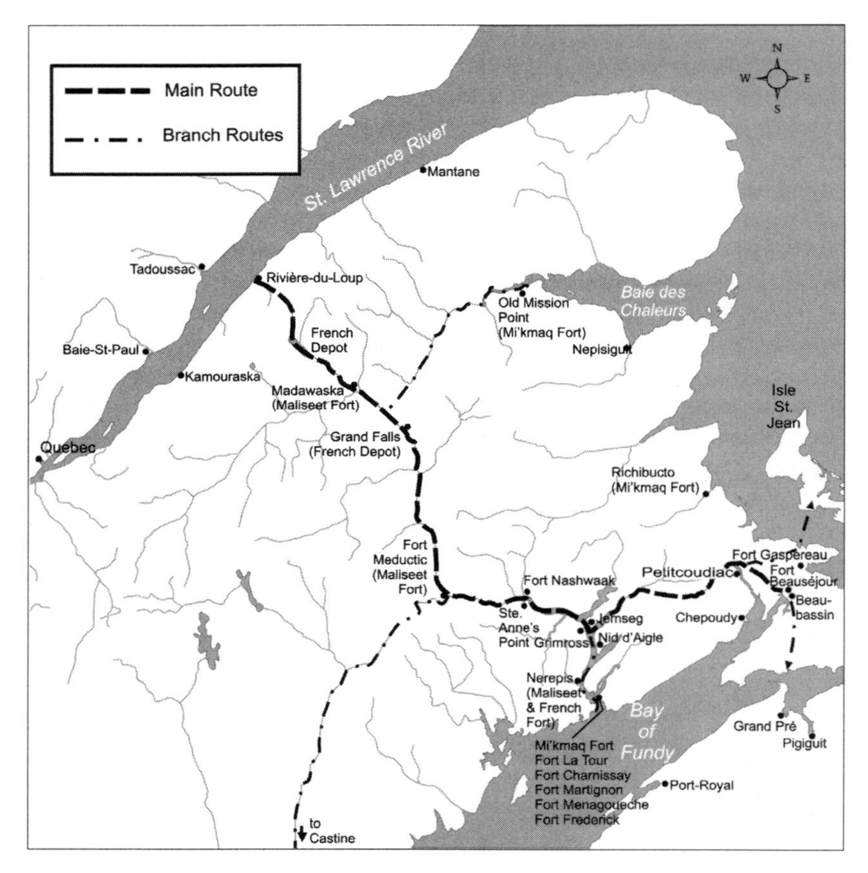

The Grand Communications Route. During the French period, the main route ran between Rivière-du-Loup and Fort Beauséjour. Secondary routes led towards Old Mission Point, on the Bay of Chaleur; the forts at the mouth of the St. John River (Saint John); Isle St. Jean (Prince Edward Island), Isle Royale (Cape Breton Island); and Port-Royal (Annapolis Royal, Nova Scotia).

Chapter One

Establishing the Route: The Beginnings to 1760

"I sent a letter to Count Frontenac by a canoe which was going to Quebec."

— Joseph Robineau de Villebon, governor of Acadia, Fort Nashwaak, September 20, 1698

The story of the Grand Communications Route begins with the end of the last ice age. When the first people arrived in what is now New Brunswick, between 6,000 and 10,000 years ago, they found a system of interlocking rivers that was marvellously suited for transportation. The central feature was the St. John River. Rising in northern Maine, the St. John flows approximately 725 kilometres to the Bay of Fundy, with few obstacles. The first, the Reversing Falls at the mouth of the river, could be negotiated at high tide or bypassed using a portage. While there were rapids at Meductic (now flooded), the only other impediment was at Grand Falls. Here, the St. John River plunges twenty-three metres and then rushes through a two-kilometre gorge. This obstruction could be bypassed only by a portage. Then, once the rapids at the mouth of the Madawaska River at Edmundston (or Little Falls) were passed, the way was clear up the Madawaska to Lake Temiscouata. From there, two series of rivers and small lakes led to Trois Pistoles on the St. Lawrence River. Later, a portage road was cut from Cabano to give a more direct route to St. André and later to Rivière-du-Loup. All told, the route from

Rivière-du-Loup to Saint John was approximately 515 kilometres, about 435 kilometres of which were by water.

In addition to providing a route from the south shore of the St. Lawrence River to the Bay of Fundy, the St. John River system has a number of branches that give access to the whole area. North of Grand Falls, the Grand River-Wagan portage route led to the Restigouche River and the Bay of Chaleur. Below Woodstock, at Meductic, an important portage route led west to the Eel River and, by a series of portages, to the Passamaquoddy, Penobscot and Machias rivers in Maine. At Oromocto, a route along the Oromocto and Magaquadavic rivers gave access to the St. Croix and rivers in Maine. A little further downstream, the Lake Washademoak-Canaan River-Petitcodiac River route led to Nova Scotia and Prince Edward Island.

The Natives of the region developed birch bark canoes about 3,000 years ago. These craft weighed only about forty-five kilograms and could carry four adults. Capable of being paddled or poled and easily carried over portages, they were ideally suited for their function. With their canoes and river systems, the first people of New Brunswick had a transportation network that was the equal of modern roads.

When French explorers arrived in the early seventeenth century, four main groups of Native people lived along this system. All were of Algonquin stock and appear to have arrived in two distinct migrations from the west. The first were the Mi'kmaq, who lived in eastern New Brunswick and Nova Scotia. It is thought that they arrived via the St. Lawrence River and the Gaspé, but parts of the Grand Communications Route may have been on their migration path. Later, the related Maliseet, Passamaquoddy and Penobscot peoples moved in from the southwest. Their tribal boundaries were well defined by the river watersheds, with the Maliseet living along the St. John River, the Passamaquoddy around Passamaquoddy Bay and the St. Croix River, and the Penobscot further south in the area of the Penobscot River in Maine.

Archaeological evidence shows that the Grand Communications Route was used for trade, hunting, fishing and war. Maliseet and Mi'kmaq legends testify to a history of warfare, but, while the First Nations of New Bruns-

Micmac Indians Poling a Canoe Up a Rapid, Oromocto Lake, New Brunswick (1835-1846), by Richard Levinge. The birch bark canoe, developed about 3,000 years ago, enabled Native people to travel easily over the interlocking systems of lakes, rivers and portages in what is now Nova Scotia, New Brunswick and Maine. NAC R9266-302

wick were fierce warriors, they do not appear to have fought amongst themselves. The common enemy of both the Maliseet and the Mi'kmaq were the Mohawk, who lived at that time along the St. Lawrence River in the area of Quebec City and Montreal, although they later withdrew into what is now upstate New York. Mohawk war parties came down the route from the north, while Maliseet and Mi'kmaq war parties travelled up it. Perhaps the best known war story tells of Malabeam, who saved her people from a Mohawk war party by luring them to their deaths over the waterfall Checanekepeag — the Destroying Giant — at Grand Falls.

Fear of the Mohawk led the Abenaki to build the first known fortifications along the Grand Communications Route at Edmundston; another fort was at the important Maliseet village of Meductic, and indications suggest that there was one at Aucpac, on Hartts Island, above Fredericton. Along the lower river, forts were located at the mouth of the Nerepis

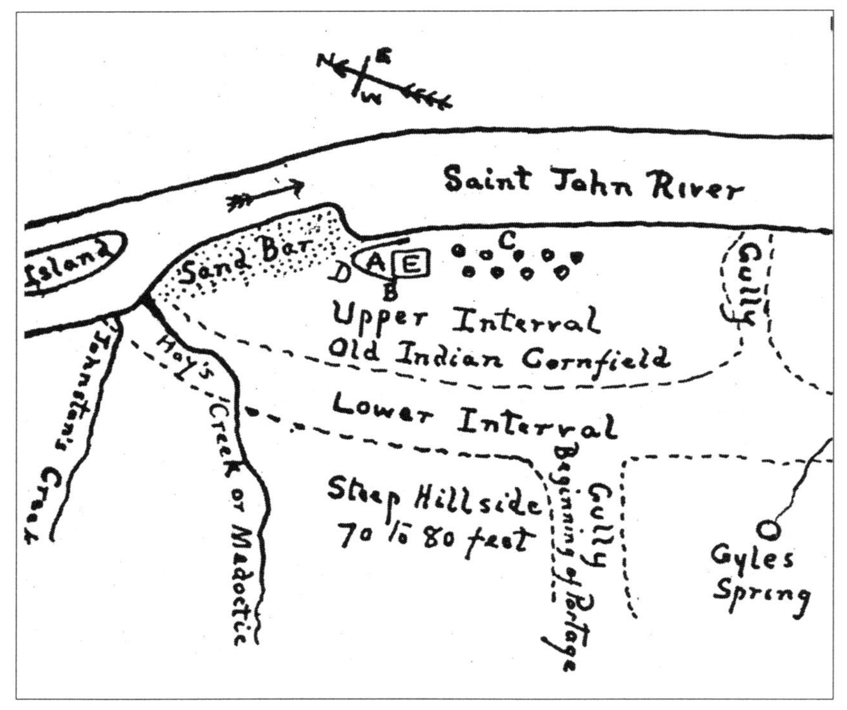

The Maliseet fort at Meductic. A: council place; B: church site; C: camping place with wigwams; D: fort site; E: graveyard. "Beginning of portage" is the head of the Eel River portage route to the Penobscot River system in what is now Maine. The site was inundated by the Mactaquac headpond in 1967. W.O. RAYMOND, *THE RIVER ST. JOHN*

River and on Navy Island, which is now under a pier of the St. John Harbour Bridge. The Mi'kmaq had forts at Richibucto and at Old Mission Point near the mouth of the Restigouche River. These defensive forts, which offered shelter to the local inhabitants when threatened by attack, were simply built but effective. Dr. W.F. Ganong describes the one at Meductic as having a central cabin surrounded by a ditch and parapet that was topped with a palisade. The fear of attackers must have been strong, for the Maliseet had only stone tools with which to dig the ditch and to cut and shape the trees for the stockade. Mohawk raids into the St. John valley continued until the 1660s.

The established patterns of life began to change in New Brunswick in the early seventeenth century. On June 24, 1604, the feast day of Saint John, a French exploring party led by Sieur Pierre Du Gua de Monts entered the Saint John harbour. Their arrival marked the start of the French colony of Acadia, which would last until 1755. Unlike the British colonial period that followed, the first hundred and fifty years of European occupation were rough and tumble, characterized by almost constant fighting. Ownership of Acadia changed hands several times, and as it changed, so did the importance of the Grand Communications Route.

After a disastrous winter on St. Croix Island, de Monts moved his settlement to Port-Royal in the Annapolis Basin. This colony, in what is now Nova Scotia, struggled on until 1613, when Captain Samuel Argall of Virginia destroyed it. In 1621, James I of Britain gave Sir William Alexander a grant that included peninsular Nova Scotia and New Brunswick, which he named Alexandria. As part of his plan to colonize the region, he sold baronetcies to Scottish gentry for £1,350. This settlement scheme faltered, and when the Treaty of St. Germain returned Acadia to France in 1632, the Sieur Isaac de Razilly led the first group of new colonists to the area.

Acadia was really a business venture of the Company of New France, similar to the Company of One Hundred Associates that controlled the colony at Quebec. Both companies were under the influence of Cardinal Richelieu. When de Razilly died in 1635, his authority as lieutenant-governor in Acadia passed to Sieur Charles de Menou d'Aulnay, who had accompanied him from France. Sieur Charles de Saint-Etienne de La Tour contested this transfer of power. Remaining in Acadia after Argall's conquest, he, too, had been named lieutenant-governor of Acadia. D'Aulnay's grants included the rich fur-producing area of the St. John River valley, at the mouth of which La Tour had built his fort. At the same time, La Tour's grant included Port-Royal, where d'Aulnay had his base. D'Aulnay and La Tour became bitter rivals, and a period of Acadian civil war ensued. The rivalry ended in 1645, when d'Aulnay captured Fort La Tour, hanged the garrison, destroyed the fort itself and built a new fort on the west side of the harbour, which he named Fort Charnissay. When d'Aulnay drowned in 1650, La Tour returned as governor; in 1653, to mend any conflicts, he married d'Aulnay's widow.

The next year, Acadia fell to the British. Sir Thomas Temple came from England to be governor; his two deputies were William Crowne, of Massachusetts, and Charles La Tour. It turned out that both La Tour and his father had been Baronets of Nova Scotia, and La Tour received a large grant of land on this basis. Temple built a fur-trading fort at Jemseg in 1659, the first European fort on the St. John River itself. Although Charles II of Britain confirmed Temple's grant in 1660 when the British monarchy was restored, the Treaty of Breda in 1667 returned Acadia once again to France. In 1663, King Louis XIV of France had abolished the commercial companies and instituted royal control over the colonies, and thus France's North American colonies became the direct preserve of the state. The great age of imperial rivalry began.

Before this time, there is little mention of the Grand Communications Route. Native people continued to use the route as they always had, and the occasional French trader, traveller or missionary probably joined them. Once the king took control and the governor or lieutenant-governor of Acadia reported to the governor-general in New France, the situation changed. Now there was a command relationship between the colonies and a requirement to exchange official correspondence. The route began to take on administrative importance.

During this period, Louis de Buade, comte de Frontenac, the governor-general in New France, began granting large tracts of land as seigneuries. One of the grantees, Martin d'Aprendestiguy, Sieur de Martignon, received land on the west side of the mouth of the St. John River. There he rebuilt Fort Charnissay and renamed it Fort Martignon. Other seigneuries granted further up the St. John River began the European settlement of the route. However, the Acadians were not to live on their land grants in peace: this was the age of piracy, buccaneering and freebootery. In 1674, a Dutch captain, Jurriaen Aernoutsz, led an expedition from the

West Indies against the British colonies. When he heard that the British and Dutch had signed a peace treaty, he turned against the French, capturing the French fort at Pentagoet (Castine, Maine). After he took Fort Jemseg on the St. John River, he allowed a French officer to go to Quebec to report the events. Count Frontenac then sent a force by canoe over the Grand Communications Route to ascertain the state of affairs and escort any French refugees to New France. This is the first known movement of European troops over the route. Having declared Acadia to be a Dutch colony and picking up a cargo of coal at Grand Lake, Aernoutsz departed. Dutch control was ineffectual, the New England colonists drove the Dutch out, and France regained control of the area.

This change of ownership had unfortunate consequences for the New England colonies. In 1687, Louis-Alexandre Desfriches, Sieur de Meneval, the newly appointed governor of Acadia, was ordered to keep foreigners — that is, British colonists — from fishing in Acadian waters and trading with the Natives, and was instructed to establish and enforce the western border of Acadia along the Kennebec River. Thus began the vicious raids known as *la petite guerre* that were intended to strengthen the French position and curb British expansion. The river system of New Brunswick was essential to organizing and mounting these raids. French-led war parties, which included Maliseet and Mi'kmaq warriors, gathered at Meductic and then followed the Eel River portage route to their forward base at Castine, Maine.

The best record of this period is the daily journal kept by Joseph Robineau de Villebon, commandant in Acadia. De Villebon arrived in Acadia in the summer of 1690, only to find that a Massachusetts expedition led by Sir William Phips had captured Port-Royal and taken away Meneval. De Villebon went to Fort Jemseg and explained the situation to the Natives and Acadian settlers there. He then went to Quebec, via the Grand Communications Route, and on to France. Phips had also captured the annual French supply ship at Port-Royal, and the presents and other goods destined for their Native allies had been lost. Now the Natives were in urgent need of powder and lead, so de

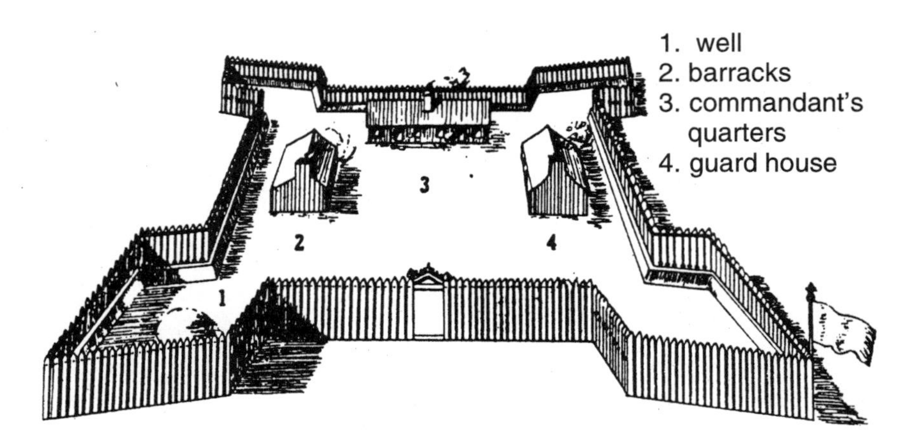

Fort Nashwaak (Fort St. Joseph), as envisioned in J.C Webster's *Acadia at the End of the Seventeenth Century*. Constructed in 1692, the fort withstood a British attack in October, 1696, and served as the capital of Acadia until 1698.

Villebon arranged for a quantity of both to be sent back to them from Quebec. This is the first recorded instance of supplies being moved over the route from Quebec to Acadia.

On his return, de Villebon moved his headquarters from Fort Jemseg to a new fort at the junction of the Nashwaak and St. John rivers, the present site of Fredericton. Fort Nashwaak, or Fort St. Joseph, was two hundred feet square, with four bastions, and was surrounded by a palisade and ditch. It was conveniently located near the large Maliseet village at Aucpac, which had eclipsed Meductic in importance. De Villebon used this site as his headquarters to launch raids against the British settlements. In 1696, following the loss of Fort William Henry at Pemaquid, Maine, the New Englanders carried the war back to the French. A force under Major Benjamin Church ravaged Acadian settlements at Chignecto and raided the new fort the French were building at Saint John. A force under Colonel John Hawthorne then joined Church, and together they moved up the St. John River to attack Fort Nashwaak. In preparation for this assault, which began on October 18, de Villebon built a second palisade around the fort and mounted ten cannon and eight swivel guns on its walls.

The British set up camp and a battery across the Nashwaak River from the fort on the east bank of the St. John River, about where the Fort Nashwaak Motel now stands. After two days of ineffectual bombardment and harassment by the French and Native skirmishers in the woods, the British departed on October 20. They burned Acadian homesteads along the way, except for that of Sieur Louis d'Amour Chauffour at Jemseg, which was spared because of a note written by John Gyles. A New England boy who had been captured by the Maliseet in 1689, Gyles was befriended by the de Chauffour family, who managed to arrange for his release two years later.

This conflict reinforced the idea that, in order to defend the settlements along the river, its mouth at Saint John had to be secured. This logic, plus the need to protect the French privateers in the area, lead to the rebuilding of the fort at Saint John. In 1698, de Villebon completed the new Fort St. Jean and made it his seat of government; he died there in 1700.

De Villebon was the first European to make strategic use of the Grand Communications Route in time of war, and he understood its importance in the struggle for North America. His journal is full of references to couriers using the route to carry messages to and from Count Frontenac at Quebec. The preferred — and indeed the only practical — resupply route was by sea, but shipping was frequently interrupted by the British. Although more than one plea was made to Frontenac to replace lost supplies, with the exception of the powder and lead sent back to Quebec in 1690, there is no mention of any supplies being carried over this route. Troop reinforcements, however, came down from Quebec and prisoners were taken back to Quebec along the route. Canoes were dispatched to Quebec at least once and sometimes twice a month, and a round trip could take as little as twenty-one days. Thus, by 1700, the Grand Communications Route had become an important strategic factor in the administration and military affairs of Acadia.

Another issue that would affect the route arose in 1700. The British disputed the French claim that the Kennebec River, in central Maine, formed the western boundary of Acadia, the boundary that Meneval had been charged with enforcing. In 1700, a boundary commission consisting of Sebastien de Villieu and Captain Frederick Southack of Boston agreed on a compromise boundary at the Pentagoet River, but this agreement was never ratified. The international boundary between New England and Acadia, and later Maine and New Brunswick, would be a vexation for the next hundred and forty-three years.

The Treaty of Ryswick ended the French and British war in 1697. De Villebon's successor, Jacques-François de Monbeton de Brouillan, moved the seat of government to Port-Royal in 1701 and razed Fort St. Jean. This decision left the Acadian settlements at Freneuse (now Sheffield) and Jemseg undefended, and when spring floods forced the settlers to abandon their land, they had to shift to other locations, such as Port-Royal. However, the French missionaries remained amongst their Native charges and continued to incite them in their continuing raids against the British settlements. Following the outbreak of war in 1702, the New Englanders sent unsuccessful expeditions against Port-Royal in 1704 and 1707. Then, in October, 1710, a combined British regular and colonial expedition finally captured Port-Royal. The Treaty of Utrecht in 1713 confirmed that it would not be returned to France. The British renamed it Annapolis Royal and called their new colony Nova Scotia.

By treaty, the French ceded "all Nova Scotia, or Acadia, comprehended within its ancient boundaries" to Britain in 1713. The British assumed this declaration included New Brunswick and Maine as far as the Kennebec River. The French disagreed and claimed they had ceded only peninsular Nova Scotia. They began enclosing the British in this area by building up their strength on the Chignecto Isthmus, Isle St. Jean (Prince Edward Island) and Isle Royale (Cape Breton Island), and in 1720 they started building a major fortification at Louisbourg. This ploy allowed the French to retain control of the Grand Communications Route and the critical access to New France for reinforcements and supplies that it provided, but it placed the Acadian settlers squarely in the middle; their loyalty, or at least their neutrality, was pulled in two directions.

After 1713, the British ruled the land lightly and had little presence outside their Annapolis Royal headquarters. They tried to gain the loyalty of the St. John River Maliseet and to make the Acadian settlers living there swear oaths of allegiance. The French strongly resisted these efforts and offered their own inducements to the Acadians to remain loyal to the French crown. The Native people in the region sided with the French during Lovewell's (or Dummer's) War from 1722 to 1725 and King George's War from 1744 to 1749. During these wars, the focus of the French-led raids changed from New England to Annapolis Royal, and forces moved between the St. John River and Chignecto, en route to Annapolis Royal, along the Washademoak -Canaan-Petitcodiac portage route; troops from Louisbourg provided additional reinforcements.

The French and Native raids on Annapolis Royal in the summer of 1744 so enraged Governor William Shirley of Massachusetts that he sent an expedition under William Pepperrell to attack Louisbourg. This siege started on April 30, 1745, and ended successfully on June 15. At the same time, a force of some six hundred French and Natives, including Huron, led by the Sieur Joseph Marin, came from Canada down the Grand Communications Route and unsuccessfully besieged Annapolis Royal. During this period, Captain William Pote of the Boston schooner *Montague* was captured, and his journal of captivity describes his journey from the Petitcodiac River to the St. John River and then onwards over the Grand Communications Route to Quebec. He and his captors obtained food and lodging at the Acadian and Native settlements along the lower St. John River, but between Edmundston and Trois Pistoles on the St. Lawrence, they fared much worse. The Huron had left a cache of food at Lake Temiscouata, but others had used it, and consequently they went hungry during the rest of their trip to Quebec. The Treaty of Aix-la-Chapelle in 1748 ended this war, and England returned Cape Breton and Louisbourg to France.

The next decade sealed the fate of New France. Although Britain and France were at peace, both sides enhanced their defences in Acadia. While France strengthened Louisbourg, Lieutenant-Colonel Edward Cornwallis founded the settlement of Halifax on June 21, 1749. With its excellent harbour, Halifax would become the British counterfoil to Louisbourg. Maliseet chiefs were brought from the St. John River to Halifax to renew the 1728 Treaty of Peace. However, there was to be no peace. The French protested any British attempt to establish a presence on the St. John River or to reaffirm the loyalties of the Acadians living there, claiming that it was still their territory. To reinforce this claim, Roland-Michel Barrin, comte de La Galissonière, the governor-general of New France, sent Sieur Charles de Champs de Boishébert with a small force to the mouth of the St. John River in the summer of 1749 with orders to rebuild the fort. When Governor Cornwallis heard of this action, he sent Captain John Rous to challenge the French presence. The two men reached a compromise: Boishébert would not refortify Saint John, but he could remain on the river until the following year, pending a resolution of the boundary question. Boishébert then moved upriver to the mouth of the Nerepis River and built Fort de Nerepice, or Fort Boishébert, as his base.

Despite British protests, the French now had undisputed control of the St. John River and the critical inland communication line from New France to the Chignecto Isthmus. The new governor of New France, Pierre-Jacques de Taffanel, marquis de La Jonquière, realized that the St. John River was the key to the country. He planned to keep the Maliseet and Mi'kmaq raiding the British settlements, secretly supplying them with arms and ammunition, and the French missionaries were key players in this scheme. This strategy was a continuation of the Mi'kmaq War against the settlement at Halifax, which had been going on since the 1720s. By limiting the British presence in Acadia, he hoped to improve the prospect of a favourable decision when Britain and France finally settled their boundary disputes.

In 1746, La Jonquière, realizing that the British could intercept French supply ships going to the Bay of Fundy, ordered a road cut along the portage route from Rivière-du-Loup to Lake Temiscouata. This track would permit pack animals and small carts to carry larger loads to Lake Temiscouata for transportation to the lower St. John River or the Chignecto Peninsula. It would also facilitate the movement of larger parties of troops and Native warriors. In 1751, he ordered Boishébert to fortify the entrance to the St. John River, which became known as Fort Menagoueche. Boishébert was soon called away to France, and his deputy, Sieur Ignace-Philippe Aubert de Gaspé, completed the task; the fort at Nerepis was abandoned.

Additional forts were under construction on the Chignecto Isthmus to secure the boundary of Acadia. In 1750, the British built Fort Lawrence in an attempt to curb the French and Native raids against Halifax. Simultaneously, the French built Fort Beauséjour on the southern end of the isthmus and Fort Gaspereau on the northern end, which effectively sealed the western border of Nova Scotia and gave the French a more secure base for raids.

The Grand Communications Route was now the strategic link between the threatened Atlantic colonies and New France. Messengers, troops and some supplies passed along the route. A force of three hundred French and Natives travelled from New France to Beauséjour, a distance of eight hundred kilometres, in less than a month. This passage was a strain on the Acadians along the St. John River because, to feed the troops, they had to slaughter draft animals and consume precious seed grain.

Abbé Jean-Louis Le Loutre, the French missionary, commented in 1753 that "it is very easy to maintain communication with Quebec, winter and summer, by the River St. John, and the route is particularly convenient for detachments of troops needed either for attack or defence." He then described the stages of the route:

> From Quebec to Rivière-du-Loup
> From Rivière-du-Loup by a portage of eighteen leagues to Lake Temiscouata
> From Lake Temiscouata to Madaoechka (Madawaska)
> From Madaoechka to Grand Falls
> From Grand Falls to Medoctek (Meductic)

From Medoctek to Ecouba (Aucpac)
From Ecouba to Jemsec (Jemseg)
From Jemsec, leaving the River St. John and traversing Dagidemoech (Washademoak) Lake, ascending by the river of the same name, thence by a portage of six leagues to the River Petkoudiak (Petitcodiac)
From Petkoudiak to Memeramcouk (Memramcook) descending the river which bears that name
From Memeramcouk by a portage of three leagues to Nechkak (Westcock, near Sackville)
From Nechkak to Beauséjour

Another record of a journey over the Grand Communications Route exists from this period. The governors of Isle Royale and New France were in the habit of sending couriers during the winter months to keep each other informed of their circumstances. In the winter of 1756, an Acadian named Gauthier undertook this journey. He left Louisbourg and travelled to Shediac, where Boishébert was leading the French resistance. From there, he portaged to the Petitcodiac and then followed the route described by Abbé Le Loutre. To facilitate travel, the French had built posts with stocks of rations at Grand Falls and Lake Temiscouata. Gauthier and his guides exchanged their snowshoes for cariole sleighs when they arrived at the St. Lawrence River and had a comparatively easy trip to Quebec City. (Interestingly, Gauthier's nephew, Louis Mercure, would provide the same service to the British during the American Revolutionary War.)

Despite the official peace between France and Britain, continued harassment of the British settlements by the French and their Native allies caused the British to plan a firm and determined response. Governor Charles Lawrence of Nova Scotia concluded that the best course of action was to destroy the French forts and establish British control over all of Acadia. He and Governor Shirley of Massachusetts planned a joint campaign for the spring of 1755. A force of New Englanders and British regulars under Lieutenant-Colonel Robert Monckton arrived at Fort Lawrence on June 2, 1755, and soon advanced on Fort Beauséjour. Expected

French reinforcements did not arrive from Louisbourg, and after a short siege Beauséjour surrendered on June 16. On June 18, the British captured Fort Gaspereau, whereupon Monckton dispatched a small force under Captain Rous to take Fort Menagoueche. Boishébert, seeing that resistance was futile, destroyed the fort as best he could and retreated upriver to Worden's or Eagle's Nest, near Belleisle Bay. There he erected a *camp volant* and constructed a small battery as a rear guard for the Acadian settlements and the Grand Communications Route. The expulsion of the Acadians from the Grand Pré and Chignecto areas began a few weeks later on July 28, 1755.

However, the French were not prepared to give up the St. John River. The Acadians on the St. John River provided shelter for many of the refugees during the winter of 1755-1756. Pierre de Rigaud, marquis de Vaudreuil, the governor of New France, wanted to retain control of the route for a number of reasons. It formed an advance guard against any British expedition up the river; it helped ensure the loyalty of the Natives; it secured his line of communication with Louisbourg; and it would reinforce French claims to the area when the next treaty was negotiated. On June 1, 1756, he wrote, "I shall not recall M. de Boishébert nor the missionaries, nor withdraw the Acadians into the heart of the colony until the last extremity, and when it shall be mortally impossible to do better." He also sent supplies to Boishébert, presumably over the Grand Communications Route, for the winter.

On July 26, 1758, Louisbourg fell to the British, and on September 18, troops under Monckton returned to Saint John. The force consisted of about two thousand men, including 350 New England Rangers, the 35th Regiment, the 2nd Battalion of the 60th (Royal American) Regiment, a train of the Royal Artillery and other support troops. They landed on September 20 and began building Fort Frederick on the site of Fort Menagoueche. Once the fort was finished, Monckton set out upriver in late October to destroy the Acadian settlements. He went as far as Maugerville before returning to Saint John, having killed or captured a number of the inhabitants; others fled into the woods or further upriver, some as far as Quebec.

A View of the Plundering and Burning of the City of Grimross (1758), drawn "on the spot" by Thomas Davies. The Acadian village of Grimross (Gagetown), identified by Davies as "the capital of the neutral settlements in the River St. John's," was burned by the British army, commanded by Lieutenant-Colonel Robert Monckton, in October, 1758, as part of Monckton's attempt to expel the Acadians from the valley. NGC 6270

After the fall of Fort Beauséjour, Boishébert was called away to lead the French resistance in the Moncton-Shediac area. His deputy, Sieur de Gaspé, withdrew up the St. John River as the British advanced. He and about twenty troops appear to have stayed for a time at Ste. Anne's Point (Fredericton) before withdrawing along the Grand Communications Route to Quebec. Wishing to disperse the Acadian settlement at Ste. Anne's Point, Monckton launched a winter raid. In late February, 1759, Captain John McCurdy's company of New England Rangers, led by Moses Hazen, reached the village. They struck with a vengeance, killing, capturing or driving off the inhabitants, killing the livestock and burning the buildings. For years the New Englanders had suffered from French-led attacks originating along the St. John River valley, and this raid gave them the chance to repay old debts. Unfortunately, most of the responsible parties, the French officers and missionaries, had already left the area, and the New Englanders' retribution fell on the Acadians, the helpless pawns in this imperial rivalry.

The dispersal of the Acadian settlements did not end French resistance on the St. John River. Groups of Acadians coalesced into settlements in secluded locations, such as near Oromocto. In August, 1759, a small force of about eighty British soldiers from Fort Frederick at Saint John conducted a raid upriver, capturing two schooners and "a great deal of plunder." This success led to a second raid in September, which the Acadians ambushed near French Lake on the Oromocto River. They killed five soldiers and wounded eight before the British withdrew. Once news of the fall of Quebec on September 18, 1759 arrived, resistance ended, and both Acadians and Maliseet travelled to Fort Frederick to make their peace. Two hundred of the Acadian refugees were deported to Halifax as prisoners. Meanwhile, the Passamaquoddy and Maliseet signed a Treaty of Peace on February 23, 1760. In 1762, Lieutenant Gilfred Studholme, who commanded the garrison at Saint John, attempted unsuccessfully to remove the remaining Acadians from the St. John River in preparation for settlement by the next wave of inhabitants, the New England planters.

By then, the period of French control of Acadia and New France had passed. During this time the Grand Communications Route had developed from simply a route for war, trade and personal travel into a strategic communications link between the centres of French power in New France, the Chignecto Isthmus and Louisbourg. The Grand Communications Route would retain these roles during the ensuing British period.

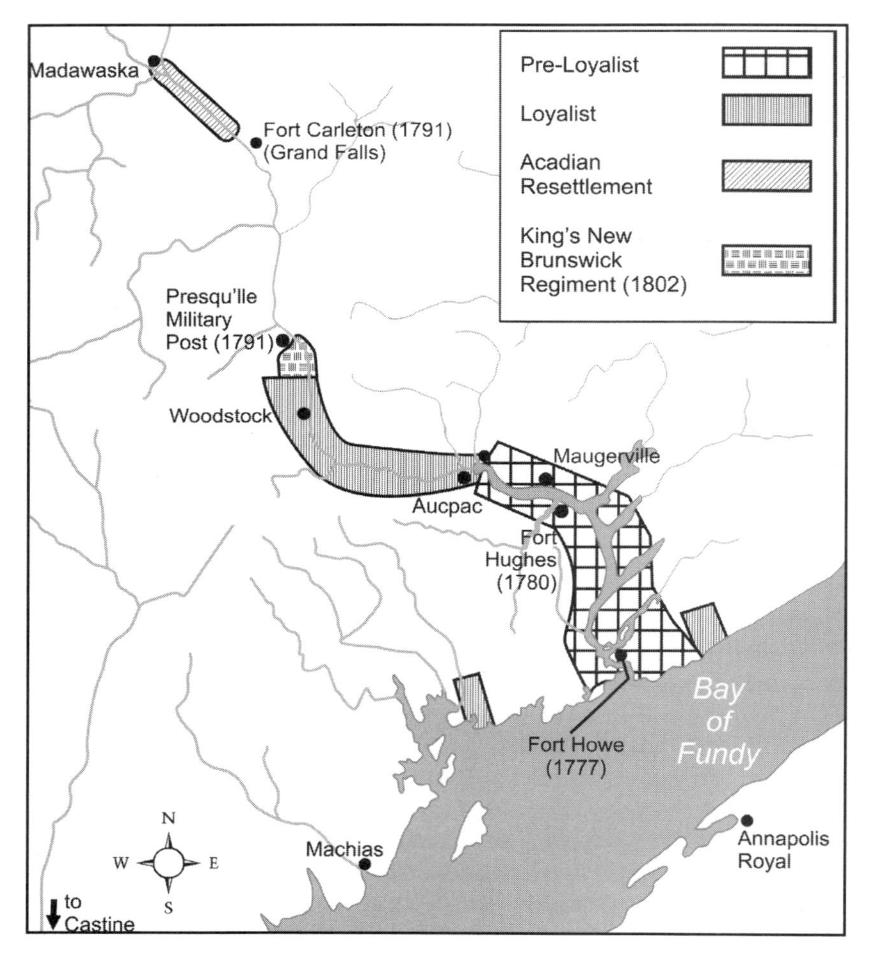

European settlement in the late eighteenth century. Following the American Revolutionary War, the British settled members of disbanded Loyalist units along the St. John River to help secure the Grand Communications Route. Security was further enhanced by the resettlement of some Acadians at Madawaska.

Chapter Two

Planters, Rebels and Loyalists: 1760 to 1785

". . . your wishes to have a commodious route from this Town [Halifax] to Quebec . . ."

— Lieutenant-Governor John Parr of Nova Scotia to Governor Frederick Haldimand of Quebec, September 2, 1783

After 1760, the Grand Communications Route briefly lost its strategic importance when the interior pathway between Canada and the Atlantic via the Richelieu River, Lake Champlain and the Hudson River could be used. This route provided easy communications to Canada during the winter when the St. Lawrence River froze. However, when the American Revolutionary War began, the Hudson passage was severed, and, once the United States achieved its independence in 1783, the Grand Communications Route regained its previous strategic significance. With the coming of the Loyalists, the British began to secure the route more permanently.

Following the expulsion of the Acadians, the British repopulated Nova Scotia with settlers from New England. Captain Francis Peabody organized a group of these planters from Essex County, Massachusetts, to settle on the St. John River. Joshua Mauger, the provincial agent in London, was instrumental in having the application approved, and the settlement, Maugerville, was named after him. In 1762, the land was surveyed and the first settlers, about 260 in all, began arriving the next year. The advance party had planned to settle in the vicinity of Ste. Anne's Point. However, a delegation of Maliseet in war paint warned them away. The planters moved twelve miles downriver, where they laid out their land grants. Other settlers soon followed, mainly officers of disbanded colonial regiments and their followers.

Despite the Treaty of Peace, the Maliseet opposed British settlement and incidents occurred. On January 20, 1765, a group of Natives who were plundering his trade goods killed Richard Simonds in Portland Point. In August, 1769, the Natives burned the house and store of Captain Godfrey Jadis, a trader living in Gage Township. There were other accounts of the Natives threatening British settlers; the years of enmity toward the British could not be easily reversed. The lieutenant-governor of Nova Scotia, Michael Francklin, became very concerned about this hostility, and, when orders arrived in 1768 to withdraw many of the garrisons, he cautioned, "If they [the Natives] should break with us before the province is better settled, it would be very difficult for the Government to prevent the destruction of the greatest part of the out settlements." Nonetheless, when British troops were needed to quell the unrest in the Thirteen Colonies caused by the enforcement of the Stamp Act in 1768, the garrison at Saint John was withdrawn and Fort Frederick was placed in the care of the trading company, Simonds and White. Consequently, on the eve of the American Revolutionary War, the Grand Communications Route was undefended.

The New York-to-Montreal communications route became one of the first casualties of the war. Soon after the rebellion started, a small force lead by Ethan Allen and Benedict Arnold captured Fort Ticonderoga on May 10, 1775, and Crown Point on the following day. These attacks effectively severed the Hudson passage. It would be some time, however, before the British turned to the St. John River route again, as two more pressing matters emerged. The first was the American invasion of Canada. By December, 1775, the Americans had captured all of Canada, except for Quebec City. Following an unsuccessful American attack on December

31, both sides waited for spring and reinforcements. When the Royal Navy arrived in May of 1776, it brought 10,000 troops, and during the campaign of 1776, the British pushed the Americans out of Canada and back to Fort Ticonderoga. Canada was now secure.

The second matter confronting the British was rebellion in Nova Scotia. The New England settlers in Nova Scotia had retained close ties with their old neighbours and shared similar political views. When the American Revolutionary War began, the centre of local rebellion was the St. John River valley. In May, 1776, the inhabitants of Maugerville formed a Committee of Safety and took control of all civil and military affairs in the district. Supported by 125 of the 137 men in the settlements, the Sunbury Committee put themselves under the protection of the Massachusetts Congress and sent a delegation there. In response to their request, congress sent muskets and ammunition that were used to arm a company of militia.

As early as the summer of 1775, Massachusetts had recommended an invasion of Nova Scotia. The Continental Congress deferred this plan because their priority was the invasion of Canada, but this decision did not deter the start of independent hostile actions. American privateers soon began raiding British commerce around Nova Scotia, sailing from Machias, Maine, a town that became a hotbed of privateering throughout the war. In August, 1775, Stephen Smith of Machias raided Saint John. He burned Fort Frederick, took its caretakers prisoner and captured a brig loaded with supplies bound for the British army at Boston. These and similar acts caused the British to reinforce the garrison at Halifax; then, on December 5, 1775, they declared martial law in Nova Scotia. They acted none too soon. The following year, the first of three proposed rebel expeditions was launched against Nova Scotia from Massachusetts, and the St. John River system was instrumental in their operations.

Early in 1776, Jonathan Eddy, a rebel from the Sackville area, met with George Washington and the Massachusetts Congress. They agreed with his plan to attack Fort Cumberland, formerly Fort Beauséjour, and authorized him to recruit an expedition. During the course of the rebel preparations, Colonel Joseph Goreham, with about three hundred men, mainly members of the Royal Fencible Americans, arrived to garrison the fort. Eddy was not deterred, and in July he left Boston, stopping at Machias to pick up more men. In early September, he arrived at Maugerville by schooner with twenty-eight men. To his delight, he found the inhabitants "almost universally to be hearty in the Cause." Captain Hugh Quinton and twenty-six men joined the rebel party. The Maliseet were divided in their support. Second Chief Ambrose St. Aubin led the faction that actively supported the Americans, while Grand Chief Pierre Thomas led the pro-British group. St. Aubin and sixteen warriors joined the Eddy expedition. Significantly, none of the Acadians along the St. John supported the rebels. Having finished his recruiting, Eddy left Maugerville on October 22. His party of about seventy-two men travelled down the St. John River and along the Fundy coast in a small fleet of canoes, whaleboats and his schooner.

Eddy's expedition met with some initial successes. On October 29, they surprised the British outpost at Shepody, and on November 8, they captured a British supply ship, the sloop *Polly*. Eddy raised some support in the Sackville area and soon laid siege to Fort Cumberland. With the sides fairly evenly matched, the siege soon became stalemated. Having only about 172 men and no artillery, Eddy could not capture the fort. Colonel Goreham, with approximately 160 men of the Royal Fencible Americans and four members of the Royal Artillery in the fort, did not have sufficient strength to attack Eddy. On November 27, the arrival of a British warship with eighty-nine Royal Marines broke the deadlock. Goreham quickly took the initiative and, on November 29, launched a successful attack that dispersed the rebels at Mount Whatley in what became known as the Camphill Rout. This victory ended any effective rebellion in the Cumberland area.

Eddy fled westward to the St. John River with a sizeable force. Besides his American and St. John River recruits, some fifty-nine men from the Cumberland area also accompanied him. Surviving accounts indicate that they likely followed the Petitcodiac-Canaan River-Lake Washademoak route to Maugerville. After resting, they continued, perhaps along the Eel River route, to Machias and then to Boston. Eddy was still in Maugerville on January 5, 1777, but likely left for Machias soon afterwards. Thus the first American invasion ended in disaster.

Meanwhile, a new and more substantial threat arose. In January, 1777, the Continental Congress appointed John Allan, another Sackville area rebel, as the superintendent of eastern Indians and colonel of infantry. He received a threefold mandate: "(1) to protect the settlements in eastern Maine; (2) to prevent the British in Nova Scotia from communicating with those in Canada through the St. John River; (3) to free Nova Scotia from British control." His first move was to prepare a second invasion of Nova Scotia. The Continental Congress authorized him to raise 3,000 men for this expedition in Massachusetts. Allan planned to secure the St. John River, capture Fort Cumberland and then continue to Halifax to burn the dockyards. Allan's grand scheme was scaled down, however, when the Americans needed forces to counter General John Burgoyne's invasion along the Richelieu-Hudson River corridor. The revised plan was reduced to a single regiment that would occupy the St. John River valley, enabling Allan to deny the British use of the Grand Communication Route and serving as a rallying point for Nova Scotia patriot refugees. He could also influence his Native allies more easily from the central location provided by the river.

While these preparations were underway, the British visited the St. John River in May. Colonel Arthur Goold and a detachment of soldiers commanded by Major Studholme arrived in response to information that Allan was attempting to establish a truckhouse. Goold continued upriver to Maugerville to deal with the rebellious inhabitants. He offered them the opportunity to renounce their rebellion and return to their allegiance to the Crown. Should they refuse this offer, Goold indicated, an armed force would return to enforce their loyalty. Following a meeting, the residents understandably agreed to accept. After administering the oath of allegiance to most of the people in the area, Goold and his party returned to Halifax. Goold also took the opportunity to meet with the Maliseet and induced several to pledge their allegiance to King George III.

Word of these events soon reached Allan at Machias. Undaunted, on May 29, he set out from Machias with a party of forty-three men. After

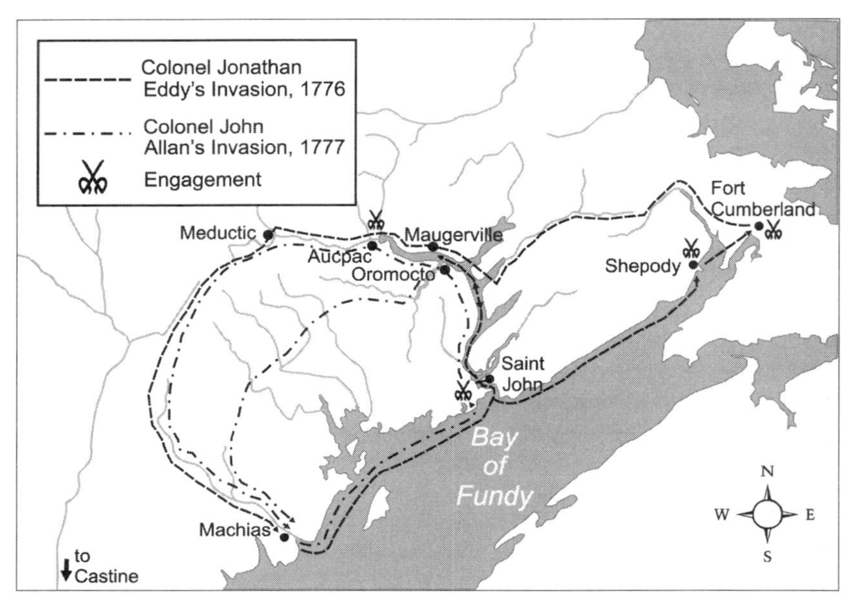

American invasions, 1776 and 1777. Both Colonel Jonathan Eddy and Colonel John Allan tried to win western Nova Scotia for the United States. They both failed and made use of the portage routes to escape to Machias, District of Maine.

stopping at Passamaquoddy for more recruits, he arrived at Saint John on June 2 and secured the area. Allan had learned that holding the mouth of the St. John River was the key to controlling the interior settlements. Leaving behind a guard of fourteen men under Captain Jabez West, he proceeded inland to Aucpac. For reasons not well explained, Allan minimized his contact with the settlers in Sunbury County and concentrated on meeting with the Maliseet to bind them to the American cause; perhaps he was sensitive to the settlers' divided loyalties and did not want to implicate them in another fiasco should he fail.

Allan also used this time to send messages to the Maliseet at Madawaska and to patriots in Cumberland, as well as communicating with his base at Machias. Additionally, he spoke with a "Frenchman" who had just come from Quebec and with another who was about to leave for there. Clearly, Allan was using his position on the St. John River as a centre from which to build an effective intelligence-gathering network based on the Grand Communications Route and its connecting routes. News travelled quickly over this intelligence system. For example, Burgoyne landed at Quebec on May 6, and by June 6 Allan had heard about it at Aucpac. Allan's ability to control this communications network and deny its use to the British was of great benefit to the Americans.

During this time, Colonel Francis Shaw arrived with forty reinforcements, but unfortunately for Allan, his regiment of 1,500 men never materialized. Even with Shaw's reinforcements he likely failed to muster more than eighty men. This force was too small to hold the river should the British return in strength, which they did. At the end of June, three British ships, the warship *Mermaid* and the sloops *Vulture* and *Hope*, with a detachment of troops commanded by Major Studholme, arrived at Saint John. On 30 June, Studholme landed with 120 men, and after a sharp skirmish, the rebels under Captain West retreated upriver. The next day, Colonel Michael Francklin, the British superintendent of Indian affairs and former lieutenant-governor of Nova Scotia, arrived with an additional 150 troops and militia. Studholme pursued the rebels towards Maugerville. After consulting with Allan, West's group fled westwards to Machias along the Oromocto-Magaquadavic portage route. The rapid British advance disquieted the pro-American faction among the Maliseet, so much so that they agreed to return to Machias with Allan. On the night of June 6, while these discussions were taking place at Mazerole's barn, Studholme surprised Allan's camp at Aucpac. Although the British did not capture any of the rebels, who were in dispersed billets, they seized their baggage, supplies, cannon and arms. This attack hastened the flight of Allan's force and a number of Maliseet — about 480 men, women and children in all — to Meductic and then over the Eel River portage route to Machias, a journey that took approximately three weeks. Thus the second American invasion also ended in disaster,

Saint John in 1815, by Joseph Brown Comingo. Constructed in 1777 to protect the St. John River settlements from invasion and raids by privateers, Fort Howe was the first step in securing the Grand Communications Route. HRSJ

and the Grand Communications Route was once more safely under British control.

While on the St. John River, Studholme and Francklin reinforced the need for continuing loyalty on the settlers and Maliseet there. However, for some reason, they did not leave behind a permanent garrison when they departed from the St. John River in July. This decision left the settlers and traders, especially those in the Saint John area who were known for their loyalty to the British, vulnerable to attacks by privateers. In early November, 1777, Captain Agreen Crabtree raided the Portland Point area and plundered the truckhouse. This and other raids prompted the traders, Hazen, White and Simmonds, to appeal to the Nova Scotia government for assistance. Relief came in late November, when Studholme returned to Saint John with fifty men of the Royal Fencible Americans, four 6-pounder cannon and a prefabricated blockhouse. Instead of reoccupying Fort Frederick, he placed the blockhouse, which he named Fort Howe, on a new site to the east overlooking the harbour. A sloop of war remained with him during the winter for protection from privateers. Soon after the fort was built, Crabtree returned to Saint John, but upon observing the state of the new defenses, he quickly withdrew.

Back in Machias, Allan continued to stir up the Natives against the British. Although the British now held the mouth of the St. John, he could still travel the portage routes, including the Eel River-to-Meductic passage, to send raiding parties to harass the British. Moreover, by using these routes and the Bay of Fundy, he was able to maintain frequent communication with the patriots and Native people throughout Nova Scotia. This contest for control of the St. John River came to a head in the summer of 1778.

The British had become concerned that Massachusetts would launch a third invasion of Nova Scotia. In response to reports that a large American force was gathering at Machias, General Eyre Massey sent reinforcements to Fort Howe in late March or early April, 1778. The threatened invasion did not materialize. Meanwhile, Allan was busy agitating among the Maliseet on the St. John. With France now allied with the Americans, Allan used the old ties of loyalty to the French king to enlist their support. As rumours of an Indian uprising spread along the valley, Francklin and his deputy for the St. John River, James White, worked to counter Allan's efforts. The Maliseet, at Allan's prodding, attacked British vessels, robbed Tory settlers and killed their cattle. During the summer of 1778, White moved along the St. John River trying to quiet the unrest. The crisis came in early September, when White intercepted a war party of ninety canoes at Long Reach. A conference ensued, during which Grand Chief Pierre Thomas declared that the Maliseet should support the British. Not all of the chiefs agreed with this pronouncement, and it was fortunate that this conference coincided with Francklin's call for a grand meeting at Fort Howe on September 24. During this conference, Francklin persuaded the Maliseet to pledge their allegiance to King George III and renounce their association with Allan and the American rebels. This diplomacy did not end Allan's influence on the St. John River, but it did effectively weaken it, although he continued to send raiding parties into the area.

Once the St. John River was secured, the British could begin using the Grand Communications Route. In October, 1778, a party of twenty-two sailors or soldiers was sent to Canada by land from Halifax. Their ship had been delayed by winds, and it was too late in the season for them to go to Quebec by sea. They were accompanied by "Thomas, Indian Chief," who was presumably their guide. By May, 1779, Allan reported that the British were "Securing every avenue, & fortifying every Necessary Post towards Canada by St. Johns, so that a constant Communication is now keep'd up." Frederick Haldimand, the governor-general of Quebec, was keenly aware of the advantages this route offered. The Grand Communications Route was his only source for news or intelligence during the winter months, "our Communication with the other parts of the Empire being in a manner totally cut off," when ice closed the St. Lawrence River to shipping.

Curiously, Haldimand's correspondence does not record any interference with his couriers by Allan or his Native allies. His greatest concern was finding reliable messengers. According to Haldimand, the Maliseet were unsatisfactory; the most dependable people were the Acadians living around Aucpac, especially Louis Mercure. Although Mercure was a lieutenant in Lieutenant-Colonel Robert Roger's Corps of Rangers (or King's American Rangers), he appears to have been employed primarily as a courier. He was also the nephew of M. Gauthier, the Acadian who had carried French dispatches from Louisbourg to Quebec in 1756. Haldimand recorded that the Acadians were clever entrepreneurs who knew they had a monopoly on the courier business. They charged and received high fees for their services. The going rate seems to have been

$100 per trip, a huge sum, but they were not averse to asking for more or conveniently forgetting about any advances that they had received. Haldimand became very concerned about reducing these costs and avoiding over-payment. Other couriers were used, such as businessmen and military officers who were making the trip for other purposes. However, these alternative couriers did not work out, and the British always went back to using the Acadians. Louis Mercure emerged as the leader of this group and even appears to have sub-contracted the courier service to others.

Meanwhile, the British were anxious to take the war to New England. In June, 1779, they captured Castine and held it until the end of the war. Intent on creating the colony of New Ireland as a buffer between Nova Scotia and Massachusetts, they saw Castine as a rallying point for Maine loyalists, a source of masts for the Royal Navy and a bargaining chip to reinforce the British claim to eastern Maine at the inevitable peace conference.

In 1780, Studholme established a forward post at the mouth of the Oromocto River, across the St. John River from Maugerville. Named Fort Hughes after the lieutenant-governor of Nova Scotia, it consisted of a simple blockhouse manned by a detachment of the Royal Fencible Americans under the command of Lieutenant Constant Connor. The post served as a visible reminder of British authority to the settlers and Natives. The memory of their rebellious tendencies had not faded completely, and the threat of further raids against the St. John River settlements was a concern. Fort Hughes effectively blocked the Oromocto-Magaquadavic route and could defend the Sunbury County settlers if required. It helped secure the large quantities of masts then being cut from among the tall white pine trees along the St. John River; the Maliseet agreed to protect the mast cutters. Fort Hughes also became the forward point of departure for the express to Quebec. Typically, dispatches from Halifax went by road to Windsor, then by boat across the Bay of Fundy to Fort Howe, and on to Fort Hughes. There, Connor engaged a courier and settled on the rate of pay for the trip. Dispatches

Fort Hughes, constructed in 1780 to serve as a forward station on the Grand Communications Route and to protect local mast cutting operations and rebuilt in 1997. COURTESY TOWN OF OROMOCTO

to and from New York City and Penobscot went to and from Fort Howe, usually by sea. It appears that the couriers made the return trip to Quebec at least monthly and sometimes more often.

The courier service over the Grand Communications Route proved its worth. Even during the navigation season it was common practice to send duplicate dispatches, one by sea and the other by land, to ensure that at least one copy arrived safely. It was over this route that, on January 6, 1782, Connor sent Haldimand his first news that, on October 19, 1781, Lord Edward Cornwallis had surrendered to the combined Franco-American army under General Washington at Yorktown, Virginia. This defeat marked the beginning of the end of the American Revolutionary War. The official end came with the Treaty of Paris in 1783.

Meanwhile, Haldimand and Lieutenant-Governor Parr of Nova Scotia exchanged letters concerning the "idea of the great utility of opening a commodious Route from this Province [Quebec] to Halifax, in order to secure a certain and speedy Communication between the two Provinces in all seasons with the Mother Country." Halidmand suggested establishing way stations at Aucpac, Grand Falls and Lake Temiscouata. Business interests had been pressing for this service since 1780, but, as Haldimand wrote to Sir Richard Hughes, the governor of Nova Scotia, he had resisted them because the increased level of traffic might "draw the attention of the Rebels to that Route and be the means of intercepting the few messengers who occasionally pass between us." These interests eagerly supported Parr and Haldimand's plan.

In time, both Haldimand and Parr took action to make the route more permanent. In the summer of 1783, Haldimand called out the militia around Rivière-du-Loup to perform *corvée* work on the portage road from the St. Lawrence River to Lake Temiscouata. By the end of the summer, the portage road was said to be passable by carriages. For his part, Parr directed Studholme to survey the route and recommend locations for post houses or way stations. The intention was to establish a regular postal system under the postmaster-general and avoid the contentious use of Native or Acadian couriers. Finally, as Brigadier-General H.E. Fox indicated in his September 20, 1783, letter to Haldimand, the settlement of the Provincial (Loyalist) regiments would "facilitate the communication between the provinces of Canada and Nova Scotia."

Following the end of the Revolutionary War, those Americans who had remained loyal to the Crown were forced into exile. Many of these Loyalists, either as civilians or as members of the Provincial Corps, came to New Brunswick. Most of the military land grants were situated along the St. John River between Maugerville and Woodstock, where fifteen Loyalist regiments settled. Because it was believed that the Treaty of Paris would not last and that another war was imminent, the Provincial Corps grants were organized in a cantonment system that would allow for quick mobilization while securing the route from Fort Hughes at least as far as Woodstock.

The Acadians living in the vicinity of Aucpac, just above Fredericton, were uncomfortable with the sudden influx of these new neighbours. While some of them, such as Louis Mercure, held clear title to their land, others had less secure tenure. The courier system had given the Acadians direct access to the recipients of the dispatches. In the fall of 1783, Louis Mercure took advantage of this connection to petition Haldimand for land grants for the Acadians in the area of Grand Falls. Haldimand agreed, partly because it would "contribute much to facilitate the communication so much to be desired between the two provinces," and, because there was uncertainty over the location of the Quebec-Nova Scotia border, he obtained Parr's agreement. It appears that Mercure, along with twenty-three other applicants and their families, sold their property and moved to the Madawaska area in 1785.

The Grand Communications Route had regained its strategic importance, and following the Treaty of Paris, the British were committed to maintaining it and expanding it into a permanent route for the postal service. Unfortunately, the ambiguous wording of the Treaty of Paris led to sixty years of controversy and crisis as Britain and the United States attempted to determine exactly where the border lay between Maine and New Brunswick.

Chapter Three

Wars and Settlement: 1785 to 1824

"By the chain of posts, thus established, the Communication with Canada is become perfectly easy and safe."
— Lieutenant-Governor Thomas Carleton to London, November 20, 1792

The settlement of the Loyalists in 1783-1784 ensured the defence of the line of communications along the lower St. John River. The security focus now shifted to the upper river valley. The threat to this part of the route came from the wording of the Treaty of Paris that described the boundary between the new United States and British North America:

> From the North West Angle of Nova Scotia, viz. That Angle which is formed by a Line drawn due North from the Source of Saint Croix River to the Highlands along said Highlands which divide those Rivers that empty themselves into the River St. Lawrence, from those which fall into the Atlantic Ocean, to the northwestern-most Head of the Connecticut River.

The map used by the negotiators at Paris was imprecise, and the difficulties of actually locating the agreed-upon boundary on the ground quickly became apparent. This border controversy, and the tensions that arose from it, would not be resolved until 1842. In the meantime, the overland road to the Canadas remained vulnerable to attack, and, like the French before them, the British were prepared to go to war to defend the route. In the event, they nearly did.

The first step in resolving the border controversy was the identification of the St. Croix River. This name was not in common use, and two rivers that flowed into Passamaquoddy Bay were the likely candidates. The Americans believed it was the Magaquadavic River (easterly river) while the British believed it was the Schoodic River (westerly river). Had the American claim been confirmed, the border would have crossed the St. John River just below Woodstock and would have effectively denied the British use of the Grand Communications Route. This controversy eventually led to the establishment of the St. Croix Commission in 1796, which determined that the Schoodic River was the St. Croix. The source of the St. Croix River was located and marked at Monument, on the Maine-New Brunswick border, but the course of the border north from there could not be determined because the "highlands" could not be located. As a result, a triangle of land comprising 12,027 square miles (3,114,993 hectares) remained in dispute. The British claimed that the border turned westward around Mars Hill and included the watershed of the Aroostook and St. John rivers. The Americans claimed the border continued due north until it reached the watershed along the south bank of the St. Lawrence River, then turned to the west. If this claim had been upheld, the Grand Communications Route would have been severed just above Grand Falls. This disagreement remained at the centre of the border controversy for the next sixty years.

In 1784, New Brunswick became a separate colony, and Colonel Thomas Carleton, the brother of Sir Guy Carleton, was appointed governor. One of Thomas Carleton's first tasks was to establish a capital for the new province. Saint John, although the major settlement within the province, was too exposed to enemy attack. After a survey, Carleton chose Ste. Anne's Point as the new capital, which he renamed Fredericton in honour of one of King George's sons. He had several good military reasons for this decision. Fredericton was at the head of navigation for many of the ocean-going ships of the day. Because it lay astride the Grand Communications Route, a strong garrison based there could quickly reinforce Saint John or move north to Canada. Troops could also be sent to the northeast via the Nashwaak-Miramichi portage route, to the southeast over the Washademoak-Canaan-Petitcodiac portage, or to the southwest along the nearby Oromocto-Magaquadavic portage. Fredericton soon developed into a major garrison town with barracks, storehouses, hospital, artillery park and powder magazine. Much of this establishment remains in the historic Military Compound.

Although the Secretary at War had promised Carleton two regiments in 1783 for the purpose of "affording protection and assistance to the Loyalists who were settling there," he had to make do with one, the 54th. They were stationed in Fredericton, Saint John and Fort Cumberland, which meant that Carleton's plan to build a battalion-sized barracks at the Falls of the Oromocto (now Fredericton Junction) did not happen. A garrison at this location could have easily moved over the portage route to St. Andrews or, via the St. John River, to Fredericton or Saint John in order to counter any threat. William Francis Ganong, in *Historic Sites of New Brunswick*, suggests that a blockhouse was built near the falls during this period, but this claim has not been confirmed.

Difficulties between the Maliseet and the new Loyalist settlers remained. In 1786-1787, Natives at Meductic threatened the settlers and attempted to kill a militia officer. In a letter to them, Carleton warned that he would not tolerate bad behaviour. Trouble between the Maliseet and Acadians also arose in the Madawaska Settlement, where the sharp practices of some traders led to strife. Carleton established a provisions and arms depot at Madawaska for use by the settlers should fighting erupt. As a result of this tension, the fledgling settlements were in danger of abandonment, and, more importantly, communication with Canada was "rendered precarious and unsafe." Carleton took firm action, demanding that a second regiment be sent to New Brunswick. In

Presqu'isle, St. John's River, July 1807 (1807), by George Heriot. The post at Presqu'Ile was built on a commanding height of land at the junction of the Presqu'Ile and St. John rivers. NAC C-012724

1790, the 6th Regiment arrived, which allowed him to secure the Grand Communications Route. He also established an independent company of militia at Madawaska. Relations with the Maliseet improved and the crisis passed without bloodshed.

In the summer of 1791, Carleton constructed military posts upriver at Presqu'Ile (just below Florenceville) and Grand Falls. These Upper Posts were designed to accommodate a company of infantry with accompanying support staff, including a commissariat officer and a barracks master. Each post had a barracks for the soldiers, quarters for the officers, a commissary storehouse and other outbuildings, as outlined on the diagram of how the Presqu'Ile post may have appeared circa 1814 (page 53). The post at Grand Falls was known as Fort Carleton. Initially, a company of the 6th Regiment garrisoned each site. Lieutenant Dugald Campbell, who was overseeing the construction, wrote, "I can assert that the troops quartered at the Posts . . . may be as comfortable as any in the Province 'bating the want of Society."

These posts served several purposes. The first was to secure the Grand Communications Route and enforce, through their presence, British claims to the area. The forts also encouraged settlement growth, acted as centres of community activity and provided support to the local civil authority. Some commanding officers at the Upper Posts also served as civil magistrates. In July of 1794, the posts proved their worth when Samuel G. Titcomb, an American surveyor in the employ of Massachusetts, struck the St. John River at Meductic while conducting a survey north from the supposed source of the St. Croix; there, he planted a stake and told the nearby settlers that this marked the boundary line. This claim alarmed them so much that some threatened to leave. Lieutenant Adam Allan, the post commander at Presqu'Ile, sent out a patrol to remove the stake and reassure the settlers.

In 1787, the plan initiated by the governors of Lower Canada and Nova Scotia to establish a postal service along the Grand Communications Route finally took form. A series of post-houses were established at intervals of between forty and forty-eight kilometre intervals. Ganong lists their approximate locations, circa 1840:

Just above Long's Creek
Just above the Nackawic
Near Fort Meductic
Just above the Munquart (Monquart Stream)
Just above Tobique (Perth-Andover)
Just above Aroostook
At Grand Falls
Near Siegas
Just below the mouth of the Madawaska
Halfway up the Madawaska on the east side
At the outlet of Lake Temiscouata, east side (Dégelis)
At Fort Ingall (entrance to the Grand Portage route)

The post to Canada operated on a fortnightly schedule during the summer and monthly during the winter. At first, the distance from Fredericton to Quebec was calculated to be 690 kilometres and took sixteen days to travel. Before the post-houses were built, the couriers spent half their nights sleeping in the woods. They used canoes in the summer, paddling or poling, depending on the direction of travel and the depth of the water. In winter, a report described the "couriers in company, who walk in snow shoes carrying with them their provisions, blankets, a fusil and the mails drawn on a hand sleigh."

The provincial government saw the need to improve the roads in New Brunswick to facilitate both military and civil travel and encourage settlement. Following a survey, construction of the Great Roads began. For the most part, these roads paralleled the old river and portage routes. The first road from Fredericton to Saint John ran along the east bank of the St. John River; the road on the west bank was built later. The road from Fredericton to St. Andrews ran south to Fredericton Junction, where it met the road from Oromocto, and then went southwest to St. Andrews (this part of the road is now abandoned). In 1799, Lieutenant Dugald Campbell was appointed to survey and start building roads on the upper St. John River. For many years, most of the roads were of dreadful quality, but they did offer an alternate means of transportation when the spring thaw or fall freeze-up made the St. John River impassable.

In 1789, the French Revolution began, and in 1793, Great Britain declared war on the French Republic. Carleton's two regiments were urgently needed elsewhere. The 65th Regiment, which had replaced the 54th in 1791, went to Halifax in April, and the 6th left for Barbados on June 8. The Upper Posts, which had been garrisoned by two companies of the 65th, were now empty. This reduced the garrison of regular troops in New Brunswick to a company of Royal Artillery based in Saint John. The British government realized that British North America could not be left defenceless and authorized the creation of six colonial regiments. Carleton received orders to raise the King's New Brunswick Regiment

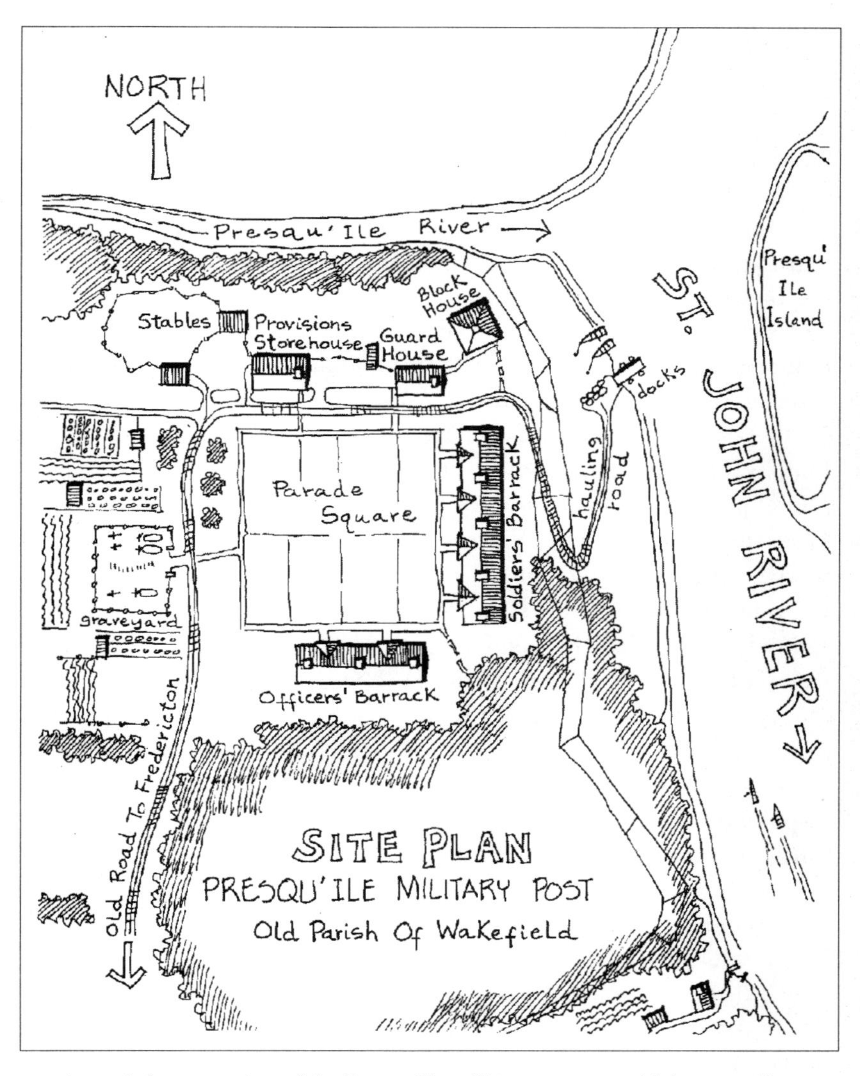

An artist's conception of the Presqu'Ile military outpost, which was well established but undefended except for the blockhouse. ERNEST A. CLARKE

(KNBR) in April, 1793. Recruitment progressed quickly, with many of the officers and men being veterans of the Loyalist corps.

Never a large unit, the KNBR was stretched to garrison St. Andrews, Saint John, Fredericton and the Upper Posts. Initially, the greatest threat to New Brunswick came from French privateers, some of which were based in friendly American ports. Some feared the United States would join in the war against Britain, which led to the repair and expansion of the Saint John fortifications and construction of three batteries at St. Andrews. A continuing concern about French spies and agents put the reduced garrisons at the Upper Posts on guard against any attempts to threaten the Grand Communications Route.

In 1794, Prince Edward was appointed the military commander in Halifax. One of his missions was to establish a line of telegraph (semaphore) posts, seven to eight miles apart, from Halifax to Fredericton. By 1800, the line ran from Halifax via Windsor to Annapolis Royal; from there, a message boat crossed the Bay of Fundy to the station on Partridge Island, and nine stations carried the line up the St. John River to Fredericton. Had the line been extended to Canada, it would have run via the Bay of Chaleur to the St. Lawrence River and then on to Quebec City. This circuitous route, designed to avoid any possible American interference, was approximately the same as the route followed by the Intercolonial Railway, built seventy years later. Apparently, this system was an effective means of communication during clear weather, but it was manpower-intensive and expensive to maintain, and it appears to have been abandoned shortly after the signing of the Treaty of Amiens in 1801.

In November, 1801, news that the Treaty of Amiens had ended the war with France reached New Brunswick. Steps were taken to call in the garrisons, including the subaltern's detachments at the Upper Posts; the KNBR disbanded on August 14, 1802. With the exception of the Royal Artillery, New Brunswick once more appears to have been without any garrison troops. The plan to establish a continual line of military settlements along the Grand Communications Route nonetheless continued. Many members of the KNBR received land grants in the new parish of Wakefield, just above the Loyalist grants.

Not surprisingly, the Treaty of Amiens did not hold, and war recommenced in 1803. On August 1, 1803, Brigadier-General Martin Hunter received authorization to raise the New Brunswick Regiment of Fencible Infantry (NBRFI). It was to consist of a Grenadier and Light Infantry Company, referred to as the Flank Companies, and eight Battalion Companies. Because of the limited pool of available men within the province, Hunter received permission to recruit outside New Brunswick. The records show the frequent use of the Grand Communications Route by recruiting parties travelling to Canada and groups of recruits coming to New Brunswick. The regiment stationed small garrisons at the Upper Posts, Fort Hughes and St. Andrews, while the bulk of the force remained in Fredericton and Saint John.

In 1807, the threat of war with the United States escalated with the Chesapeake Affair. On June 28, HMS *Leopard* forced USS *Chesapeake* to submit to a search for deserters. The means of persuasion involved gunfire that left the American ship disabled; the United States was outraged. Tensions rose along the border and many anticipated war. In response, a thousand militia assembled at Fredericton, Saint John and St. Andrews between January and April,1808, to bolster the defences of the province. The St. Andrews garrison also provided security for the Oromocto-Magaquadavic portage, while a company at Meductic guarded the Eel River portage. Construction of Fort Tipperary (which still stands in downtown St. Andrews) strengthened the defences of the town. In April, 1808, the militia were relieved when part of the 101st Regiment arrived from Halifax. In the reallocation of troops, two companies of the New Brunswick Regiment left the province, bound for Sydney, Cape Breton, and Charlottetown, Prince Edward Island. In 1810, the NBRFI's offer to volunteer for duty overseas was accepted, and they became a regular British army unit, now renamed the 104th Regiment of Foot.

The international climate continued to deteriorate, and on June 18,

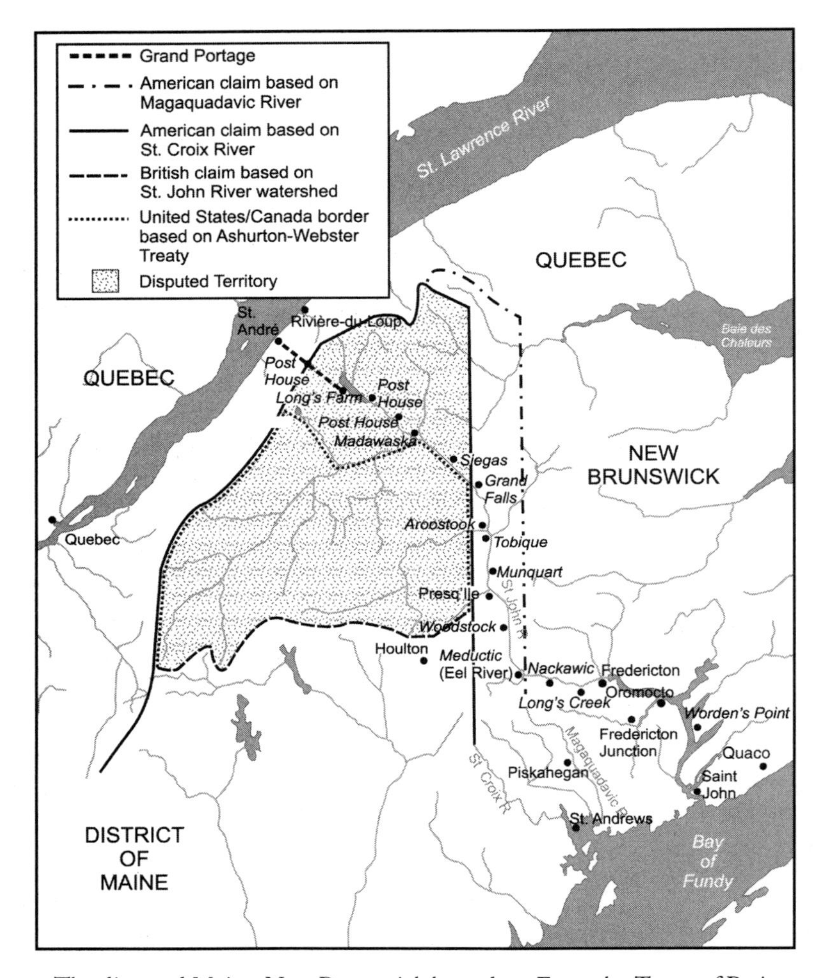

The disputed Maine-New Brunswick boundary. From the Treaty of Paris in 1783 until the ratification of the Webster-Ashburton Treaty in 1843, a poor understanding of geography created frequent conflict. Following the American Revolutionary War, the British established postal stations and built the Upper Posts to secure the Grand Communications Route. During the War of 1812, they built additional fortifications to secure this route and the Oromocto-to-St. Andrews road.

1812, the War of 1812 started. News of the war reached New Brunswick eight days later. Once more, preparations to defend the province were undertaken. Blockhouses or batteries were built or improved in St. Andrews and Saint John. The garrisons were reinforced there and at Fort Cumberland by drawing from the general reserve in Fredericton. Token forces continued to occupy the Upper Posts: four men at Grand Falls, nine at Presqu'Ile and nine at the new post at Eel River (Meductic). Soldiers stationed along the Fredericton-Saint John road expedited the movement of dispatches. New blockhouses, proposed as early as 1784, were built along the Fredericton-St. Andrews road at Fredericton Junction and Piskahegan (presently near Pomeroy Bridge on Highway 770). These posts provided security for the Great Roads and aided in the apprehension of British deserters. The St. John River defences were improved with the construction of a battery of three 10-pounders and a blockhouse at Worden's Point, about thirty miles above Saint John, and repair or rebuilding of the blockhouse at Oromocto. Seventeen armed gunboats helped defend the St. John River. A battery was also built at Quaco to protect the shipyards at St. Martins.

In North America, the war was fought both at sea and on land, the latter mainly in Upper Canada. During the 1812 campaign season in Canada, the outnumbered British forces more than held their own, but Sir George Prevost, the governor-general, anticipated a major American invasion in 1813. He desperately needed more troops, but the bulk of the British army was committed to the European war in Spain and Portugal. However, New Brunswick and Nova Scotia had reached a "non-aggression" agreement with the New England states, and some of the troops stationed there were available for service. In January, 1813, Lieutenant-General Sir John Sherbrooke, lieutenant-governor of Nova Scotia and commander of His Majesty's Troops in Nova Scotia and New Brunswick, received orders directing the 104th Regiment and a detachment of Royal Artillery to move to Canada. They would be replaced by troops from Halifax. A similar need arose for sailors to man the ships of the Provincial Marine on the Great Lakes, and a party of naval officers and seamen followed the 104th to Canada. Although the official orders were not issued until February 5, 1813, the 104th, anticipating their departure, had concentrated outlying detachments at Fredericton and begun practicing drills and marches on snowshoes. According to a local tradition, planning for the march took place in the Odell House, which still stands in Fredericton. The boys of the regiment and those men not fit to make the march remained with the two companies left in New Brunswick. Two other companies remained in Prince Edward Island and Cape Breton. In the spring, the two New Brunswick companies, along with the heavy baggage and families, proceeded to Quebec by sea, as did the Royal Artillery detachment, which could not be readied in time for the winter march.

On February 16, the 104th began the epic march from Fredericton. Each man's winter kit consisted of a greatcoat, snowshoes, moccasins, mittens and a blanket. They drew their equipment and provisions on a toboggan shared by two men. The officers carried their kit in knapsacks, or used dogs to carry equipment. The headquarters and Grenadier Company left first, with the four Battalion Companies following on successive days and the Light Company departing on February 21. Each company group formed a line about half a mile long as the men marched in single file. For the first seven days they passed through the more settled areas; marching on winter roads was not too difficult, and they could sleep in houses or barns. Once beyond Woodstock, they passed through no settlements except the barracks at Presqu'Ile and Grand Falls and the settlement at Madawaska, and they had to rough it in the woods. Each day, they started at first light and stopped in the early afternoon to cook their dinner and construct huts to sleep in. Snow fell continually, and each company faced the fatiguing task of breaking trail every day. Lieutenant John Le Couteur described the technique they used:

> In order to relieve the men, each officer and man took his turn to break the road, as it was called, by marching as leader for ten or fifteen minutes, then stepping one pace aside and letting the whole company pass him, when he

> threw off his snow-shoes and marched on a firm, hard path in the rear. It must be seen by this arrangement the first pair of snow-shoes had to break a path in front, the second pair improved the track of the first, the third and every succeeding rendered it firmer and harder, till the Toboggans came which travelled on a pretty solid path.

The cold was intense, the temperature averaging between - 27°C and - 32°C. Le Couteur described the hardship of marching in this weather, which on one occasion brought his company to a stop:

> Knowing the dangerous consequences that might ensue from a prolonged halt in such excessive cold, I hastened in the deep snow to the head of the company and, going along, I observed that almost every man was already more or less frost-bitten and was occupied in rubbing his cheeks or nose, or both, with snow. In my progress I also was caught by the nose and, when I turned the corner in the river, I really thought I should not have been able to proceed, the cold wind appeared to penetrate through my body in defiance of flannels or furs. I however urged the men on, as soon as we had taken time to lay one poor fellow on a Toboggan whose whole body was frost-bitten, and covered him with blankets. By changing the leading file every four or five minutes we at length got to the huts, having about 90 men out of 105 more or less frost-bitten on that occasion.

The troops met with a warm reception at Madawaska: the inhabitants turned out to carry them twenty-one miles by sleigh, which the weary troops much appreciated. Once they reached St. André on the St. Lawrence, the road to Quebec was well packed and they could make good time. On average, the companies took twenty-four days to complete the 564 kilometre march.

A typical winter march along the Grand Communications Route. *Travel on the St. John River on the way to Quebec in January or February.* (1815), by Lieutenant E.E. Vidal. NBM W6798

The only serious difficulty they encountered was a blizzard which prevented the last two companies from crossing Lake Temiscouata and caused them to run out of food. Lieutenant Charles Rainsford, who was familiar with the area, volunteered to go for help to St. André. There, he met William Anderson, who had been contracted by the commissariat to provide rations. The rescue party soon met the two companies, which had managed to cross the lake at the entrance to the Grand Portage, now known as Cabano.

Despite the harsh weather, everyone completed the march. Only one soldier suffered severe frostbite — Le Couteur wrote that he was, "quite a hideous spectacle, altogether one ulcerated mass, as if scalded all over from boiling water" — but he caught up with the regiment once he had recovered. On March 15, the last company of the 104th reached Quebec. After a brief rest, they set out on March 27 for Kingston, another 564 kilometres to the west. Altogether they successfully travelled over 1,128 kilometres during the dead of winter in fifty-two days, including the twelve to fourteen days' rest in Quebec. Their singular feat has never been equalled.

The movement of the 104th validated the British policy of maintaining the Grand Communications Route as a strategic means of travel between the Maritimes and Canada, especially during the winter months. In the summer of 1813, the 104th participated in the fighting along the Niagara frontier and then went into winter quarters in Kingston. In 1814, the main body of the regiment remained there on garrison duty while the Flank Companies returned to Niagara, where they served with distinction. The companies in Prince Edward Island and Cape Breton joined the regiment in Quebec in 1814.

The departure of the 104th left New Brunswick short of troops. In June, 1813, the 2nd Battalion of the 8th Regiment arrived in New Brunswick from Halifax, and a new regiment, the New Brunswick Fencibles, was raised. The Fencibles performed garrison duties at Fredericton and the Upper Posts, and several members of the unit manned gunboats on the St. John River.

The Grand Communications Route demonstrated its strategic importance again in 1814, when the need for more troops in Canada, as well as sailors for the naval war on the Great Lakes, intensified. On February 25, 1814, approximately five companies of the 2nd Battalion of the 8th Regiment left Fredericton for Canada. A detachment of 217 sailors and Royal Marines followed soon afterwards. Their march appears to have been less eventful than that of the 104th. The citizens of Saint John transported the sailors to Fredericton by sleigh, and the House of Assembly voted £300 to hire sleighs for the soldiers and sailors to ease the first part of their journey from Fredericton.

While these forces moved along the route with comparative ease, the security of the mail carried along the Grand Communications Route was a major concern. The postmaster general of Canada, James Heriot, requested a military escort of two soldiers for the mail couriers travelling between Madawaska and the Grand Portage at Lake Temiscouata. In the fall of 1813, Major-General Sir Thomas Saumarez, the commanding officer in New Brunswick, learned of an attempt earlier in the year by a suspected American agent to entice a private of the New Brunswick Fencibles to desert and bring along his bag of mail. The soldier had replaced the usual postman, who had been sent to Quebec with urgent dispatches from England. Saumarez alerted magistrates and militia officers along the St. John River, but the warning came too late to be of use. In the summer of 1814, an American tried to bribe a Native to intercept the mail going from Presqu'Ile to Canada. Nothing came of this episode because the loyal Native reported it. This incident was sufficiently important for Saumarez to travel to Presqu'Ile to personally investigate it, for news of such attempts was important enough to be brought to the attention of the senior commanders.

The threat of interference with the route was so serious that the British were prepared to go to great lengths to ensure its security. With the end of the war in Europe, Great Britain sent reinforcements to North America for the 1814 summer campaign season. The British planned four major advances, including one against the District of Maine. While all of the reasons for attacking Maine are not known, one is certainly clear. In a letter dated July 15, 1814, to Sherbrooke, Prevost wrote, "I confess I do not perfectly comprehend how you are to occupy so much of the District of Maine as shall secure an uninterrupted intercourse between Halifax and Quebec, unless it is by placing a Garrison at the Madawaska Village and extending posts from there to the Temiscouata Lake and adding to the force at the Grand Falls." The security of the Grand Communications Route was one of the objectives of the invasion. In July, Eastport, Maine, and Fort Sullivan were captured, and in August, Sherbrooke sailed from Halifax with a squadron of nine warships and ten transports carrying men of the 29th Regiment, 1st Battalion 62nd Regiment, 7th Battalion 60th Regiment and 98th Regiment plus artillery and engineer detachments. They captured Castine and Bangor without serious opposition and took Machias a few days later. The customs revenues seized at Castine were used to found Dalhousie University and the Cambridge Military Library, both of which still exist in Halifax.

On December 24, 1814, the War of 1812 ended with the Treaty of Ghent. Even though the New Brunswick House of Assembly had petitioned the Prince Regent to "direct such measures . . . as he may think proper to alter the boundaries between those States and the Province, so as that important line of communication between this and the neighbouring Province of Lower Canada, by the River Saint John, may not be interrupted," this request went unheeded. Because the Royal Navy had lost the contest on the Great Lakes, the Duke of Wellington believed the British lacked a strong bargaining position and recommended that no changes in any boundary should be requested. Thus the treaty confirmed the *status quo ante bellum*, but it did provide for two boundary commissions to determine the border north of Monument and in relation to islands in the Bay of Fundy. The first commission failed; the second succeeded.

Even before the war ended, the British moved to consolidate their control of the Grand Communications Route through land settlements. In the summer of 1814, Colonel Joseph Bouchette, the surveyor-general of Lower Canada, settled members of the 10th Royal Veterans Regiment along the Grand Portage and as far south as Grand Falls "for the purpose of facilitating the Communication between Lower Canada and New Brunswick." The general reduction of forces after the war provided the opportunity to complete the band of settlement along the Grand Communications Route. In 1817, the president and council of New Brunswick declared the land between Presqu'Ile and Grand Falls to be a Military Settlement. The Upper Posts were to become administrative centres for the new settlers, issuing rations for up to three years, farm tools, seeds, and supplies such as building hardware and blankets needed by settlers to develop their farmsteads. The first unit to be settled was the New Brunswick Fencibles. Disbanded on February 24, 1816, they were authorized land grants in the spring of 1817. Members of the 104th who disbanded at Montreal in May, 1817, followed them. The 98th regiment helped to transport members of the 104th to their new homes. They, in turn, were disbanded in the spring of 1818 and received grants "above Presqu'Ile." Their replacement regiment, the 74th, provided transport upriver for the disbanded Royal West Indian Rangers in 1819. Although a rough road had been constructed on both sides of the river as far north as Presqu'Ile, the preferred method of transportation was by riverboat, and the movement of supplies to the Military Settlement helped to foster a busy shipping industry on the river.

Not all members of these disbanded regiments received land grants in the Military Settlement, nor did they all remain on their land. However, enough stayed to create a strip of settlement along the Grand Communications Route. Members of other regiments, including the 8th and 4th Royal Veterans Battalions and perhaps the Royal Artillery, also received grants in the Military Settlement. In total, 238 soldiers and their families took up land grants before the period of assisted settlement ended in 1823. The posts at Presqu'Ile and Grand Falls were both vacated within the next year, although family members of those who had been stationed at the posts and were now settlers continued to occupy the buildings.

After 1814, the Grand Communications Route remained important. In early February, 1815, Commodore Sir Edward W.C.R. Owen arrived at Halifax en route to the Great Lakes, where he was to take command of the Provincial Marine. He and his party, including Lieutenant E.E. Vidal, made the overland march to Canada in February and March. Then, in June, 1816, a recruiting detachment of the 104th Regiment, consisting of eighty-three members of all ranks, marched from New Brunswick to Quebec as part of the consolidation of the regiment prior to its disbanding.

The peace and tranquillity that had settled over the Grand Communications Route after 1814 was not to last. The boundary commission investigating the border north of Monument faltered, and the creation of the state of Maine in 1820 added a new participant to the controversy. Maine politicians remained adamant: they would extend the boundary claim to the fullest extent. Increasing tension became a fact of life along the border for the next twenty years.

Chapter Four

Border Crises and Resolution: 1824 to 1845

"Eventually we should be compelled to make a stand when our lines of communication were interrupted."

— *Major-General Sir Archibald Campbell, lieutenant-governor of New Brunswick, to Lord Goderich, Colonial Office, October 4, 1831*

Between 1824 and 1845, a series of border crises threatened the very existence of the Grand Communications Route, and the British risked war to defend it. The politicians of the new state of Maine, stinging from the humiliation of two British invasions in the previous forty years, pressed for the most favourable claim possible under the Treaty of Paris. The political and economic issues of state's rights and control of the rich timberlands in Aroostook County also affected the boundary issue. These tensions culminated in the so-called Aroostook War and the final settlement of the border in 1845.

Starting about 1817, a small group of Americans, possibly sent as *agents provocateurs*, settled along the St. John River above the Madawaska Settlement. John Baker soon emerged as the outspoken advocate of this pro-American faction. On July 4, 1827, he and his neighbours erected a Liberty Pole at Baker Brook and raised a flag with American symbols. They then signed a petition proclaiming self-government. The same summer, Baker interfered with the passage of a mail courier on the

Madawaska River and strongly suggested to a militia officer, Captain Simon Hebert, that he not drill his militia company. In September, the British authorities arrested Baker and two associates, charging Baker with sedition. Following his trial in May, 1828, he was sentenced to two months in jail and fined £25. The fine was apparently paid by the state of Maine. This confrontational behaviour was not confined to the Madawaska area. In October, 1827, some Americans living along the Aroostook River prevented a British official from enforcing a court writ. The official later returned with reinforcements and carried out his duty.

Lumbermen from both sides of the border cut timber in the disputed area, sometimes with permits from Maine or Massachusetts or New Brunswick. The government of New Brunswick, aware of complaints by Maine that "trespass" timber was being cut illegally, attempted to quiet the tension by banning lumbering in the area, which they now referred to as the Disputed Territory. To enforce this ban, James A. MacLauchlin was appointed Warden of the Disputed Territory in 1827. Any revenues received from the sale of seized timber were placed in the Disputed Territory Fund, which would be divided once the boundary was finally settled.

In 1826, the boundary commission stalled, so the issue was submitted to an arbitrator, King William I of the Netherlands. When King William rendered his decision in January, 1831, he awarded sixty-six percent of the land to the United States, while Britain retained control of the vital Communications Route. The British accepted the arbitration, but the United States, at the insistence of Maine, rejected it, setting the scene for a jurisdictional struggle between Maine and New Brunswick. The Americans had seen how the British used roads and military posts to secure their hold on the land and decided to emulate them. In 1828, they extended the military road from Bangor to Houlton and constructed Hancock Barracks for four companies of the 2nd United States Infantry, whose presence posed a threat to the security of the Grand Communications Route. To avoid this threat, the British proposed building a military road overland from Fredericton to Grand Falls east of the river. Work commenced on the Royal Road in 1832.

Following rejection of the arbitration, Maine initiated a series of challenges to British authority. In 1831, Maine commissioned John E. Dean and Edward Kavanagh to conduct a census of the area, and these men gulled many of the inhabitants of the Madawaska Settlement into cooperating by threatening to revoke their land tenure if they did not participate. In the same year, Maine incorporated the Madawaska Settlement, which extended along both sides of the St. John River, as a town, and then organized two meetings to elect town and state officials. The British protested these challenges to their authority to the United States government. While sympathetic, the United States was not prepared to curb Maine's activities. In response, the lieutenant-governor of New Brunswick, Sir Archibald Campbell, visited the Madawaska Settlement and had the American ringleaders arrested.

In 1837, another census initiated a new crisis. The United States government returned surplus money to the states, which in turn refunded the money to their citizens, based on a census. Ebenezer S. Greely was hired by Maine to conduct the census of Madawaska. In return for stating that they were under American jurisdiction, the heads of families were promised up to twelve shillings and sixpence. Although arrested and jailed twice by the British, Greely kept returning to his task. After his third arrest, Greely said he would continue with the census supported, if necessary, by "armed forces."

Sir John Harvey, the new lieutenant-governor of New Brunswick and an experienced soldier, realized it was time to take firm action. In a letter, he told Robert Dunlap, the governor of Maine, that he held "positive instructions from [his] government not to suffer any act of sovereignty or jurisdiction to be exercised by any foreign Power within the territory in dispute" until the boundary negotiations had been concluded. Further, he was acting in accordance with the wishes of the United States government, and, if necessary, he would use force to repel "any act of invasion or foreign jurisdiction." He added that he would have agreed to the census had Maine first asked for permission to conduct it. To give substance to his words, Harvey sent two companies of the 43rd Regiment to Woodstock and Grand Falls. They left Fredericton on September 15 and travelled upriver in horse-drawn towboats. They reached Woodstock the following day, and the company proceeding to Grand Falls arrived there four days later. This force of 110 soldiers brought five hundred muskets for the volunteer militia from Woodstock, Grand Falls and the intervening area who stood ready to assist the civil authorities. The message was received in Maine and the crisis passed.

Just as the Greely uproar subsided, another one was building in the Canadas. In November and December, 1837, discontent over responsible government erupted into open rebellion in Lower Canada when the *Patriotes,* led by Louis-Joseph Papineau, clashed with British troops at St. Denis, St. Charles and St. Eustache. When Sir Francis Bond Head, the lieutenant-governor of Upper Canada, sent his only battalion, the 24th, to Lower Canada, he left Upper Canada undefended, which paved the way for William Lyon Mackenzie to begin his revolt at Toronto on December 4, 1837. The local militia quickly put down this uprising. Later in December, members of the Canadian militia outraged the Americans when they made an unauthorized raid into New York state and seized and burned the supply boat *Caroline*, which was being used by Mackenzie's rebels.

Harvey suspected that Greely's shenanigans and the threatened border conflict may have been "intended as a diversion in favour of the Papineau party, by preventing reinforcement of troops being sent from Nova Scotia." As it turned out, Maine did not take advantage of the rebellion to create any problems within the Disputed Territory. However, as the crisis in Lower Canada developed, Halifax received a request for reinforcements of at least two regiments. Thus began a massive reinforcement of Canada, which increased its troop strength from 3,150 to 10,271 within a year. While most of the troops arrived in the spring of 1838, after the opening of the navigation season, the only way to send urgently needed reinforcements in late 1837 was over the Grand Communications Route through New Brunswick.

The first regiment to move, the 43rd, stationed in New Brunswick, had to wait until freeze-up and the opening of the winter road to Quebec. Meanwhile, Harvey directed James MacLauchlin, Warden of

Scene on the River St. John, 1st Day's March (1837-1838), by Captain Godfrey Charles Mundy. The 43rd Regiment left Fredericton for Quebec in December, 1837. NAC C-117468

the Disputed Territory, to cut a rough road along the west bank of the Madawaska River and construct huts, or cabanos, at the overnight stopping places where civilian buildings were not available; sites included Dégelis and Cabano, Quebec. The 43rd left Fredericton between December 11 and 15 and arrived in Quebec by January 1, 1838. During this time, the 85th Regiment left Halifax, arrived in Saint John and began its march to Canada. They travelled about two days behind the 43rd, passing through Fredericton between December 17 and 20. The two regiments took with them two 12-pounder carronades and a cohorn mortar, mounted on sleighs, to sweep away any opposition they might encounter while moving up the St. Lawrence River to Quebec. In January, 1838, the 34th Regiment and 8th Company of the 4th Battalion, Royal Artillery, followed them upriver. All told, almost fifteen hundred men travelled along the Grand Communications Route between December 11, 1837, and the end of January, 1838.

Unlike the 104th Regiment in 1813, these troops moved all the way by sleigh. The trip from Fredericton to Quebec City took approximately thirteen days, with overnight stops arranged by the commissariat at about forty-kilometre intervals. Soldiers rode in two-horse sleighs in New Brunswick, and single-horse cariole sleighs were used beyond

Cabano, where the portage road was narrow. When the 85th crossed the Aroostook River, they had to use boats, as the ice was unsafe. The troops, many of whom had recently arrived from the tropics, were kitted out with moccasins, fur caps, mitts and extra blankets to help them cope with the temperatures, which ranged from -20°C to -32°C. According to contemporary accounts, the soldiers travelled well. Their journey captured the attention of the Duke of Wellington, who is said to have called the march of the 43rd Regiment "one of the greatest feats ever performed, and the only military achievement performed by a British officer that he really envied." This was high praise indeed.

Because of the recent tensions on the border, Harvey worried about the American response to the British troop movements through the Disputed Territory. In a letter to Lord Gosford, Harvey wrote, "I consider the right to such a passage [of troops to Canada, as was done during the late war] for our troops to be as undisputed as for our couriers." Harvey also wrote to H.S. Fox, the British Envoy Extraordinary and Minister Plenipotentiary in Washington, asking if the United States government would object to this activity. While Washington appears not to have raised any objections, the Maine legislature considered it to be "a palpable outrage upon the sovereignty of Maine, and of the United States, and a fresh cause of complaint." Nevertheless, the troop movement went ahead without any interference.

Political affairs in Upper and Lower Canada remained tense throughout 1838. Many of the rebels fled to the United States, where they found active support from anti-British sympathizers who called themselves Patriot Hunters. The hunters were organized into lodges in most of the northern states, up to ninety-nine of them in Maine. On November 4, 1838, the rebellions again broke out in Lower Canada, with a *Patriote* uprising at Beauharnois followed by raids across the border from the United States at Lacolle and Odelltown on November 7 and 9. In Upper Canada, a group of Hunters attacked Prescott and fought the Battle of the Windmill on November 16-17, 1838. Even though six regiments of infantry and two regiments of cavalry had reinforced Canada during the summer, the call again went out for reinforcements. Once more, the

Nova Scotia command (present-day New Brunswick, Nova Scotia and Prince Edward Island) was the closest source. The 65th regiment left Fredericton in mid-November, marched to Shediac, and boarded a ship for Quebec. The 11th Regiment moved forward from Halifax and passed through New Brunswick for Quebec via the land route in late December. A company from each of the 65th and 95th Regiments and the 1st Company of the 4th Battalion, Royal Artillery followed behind them. In total, 776 troops moved over the Grand Communications Route in December 1838 and January 1839.

Meanwhile, the ongoing border controversy entered its most serious phase. Maine continued to chafe at the apparent "understanding" between Washington and London concerning the interim jurisdiction exercised by Great Britain in the Disputed Territory. In response to a report of large-scale "trespass" timber cutting on the Aroostook and Fish rivers, the Maine legislature passed a resolve on January 24, 1839, authorizing Rufus McIntire, the Maine land agent, to "arrest, detain and imprison all persons found trespassing on the territory of this State" and voted $10,000 to fund the operation. The Aroostook War was about to begin.

McIntire and Hastings Strickland, the sheriff of Penobscot County, set off for the Aroostook soon afterward with a "civil posse" of about two hundred men. Advancing from their forward base at Marsardis, they encountered a group of approximately three hundred trespassing lumbermen. When the trespassers saw that the posse was armed with a brass 3-pounder cannon, they prudently withdrew. Elements of the posse continued to a place just west of present-day Fort Fairfield, capturing or driving off any trespassers they encountered along the way. McIntire and a few others carried on to Fort Fairfield, where they were captured by a group of lumbermen and carried off to the Fredericton jail. In retaliation, the posse arrested James MacLauchlin and Captain Benjamin Tibbetts when they visited the camp at Marsardis on February 16, and sent them off to Bangor. About a week later both groups were released.

When he heard about these events, Lieutenant-Governor Harvey tried to defuse the crisis by telling Maine's Governor, John Fairfield, that, if the posse withdrew, Harvey would take measures to protect the

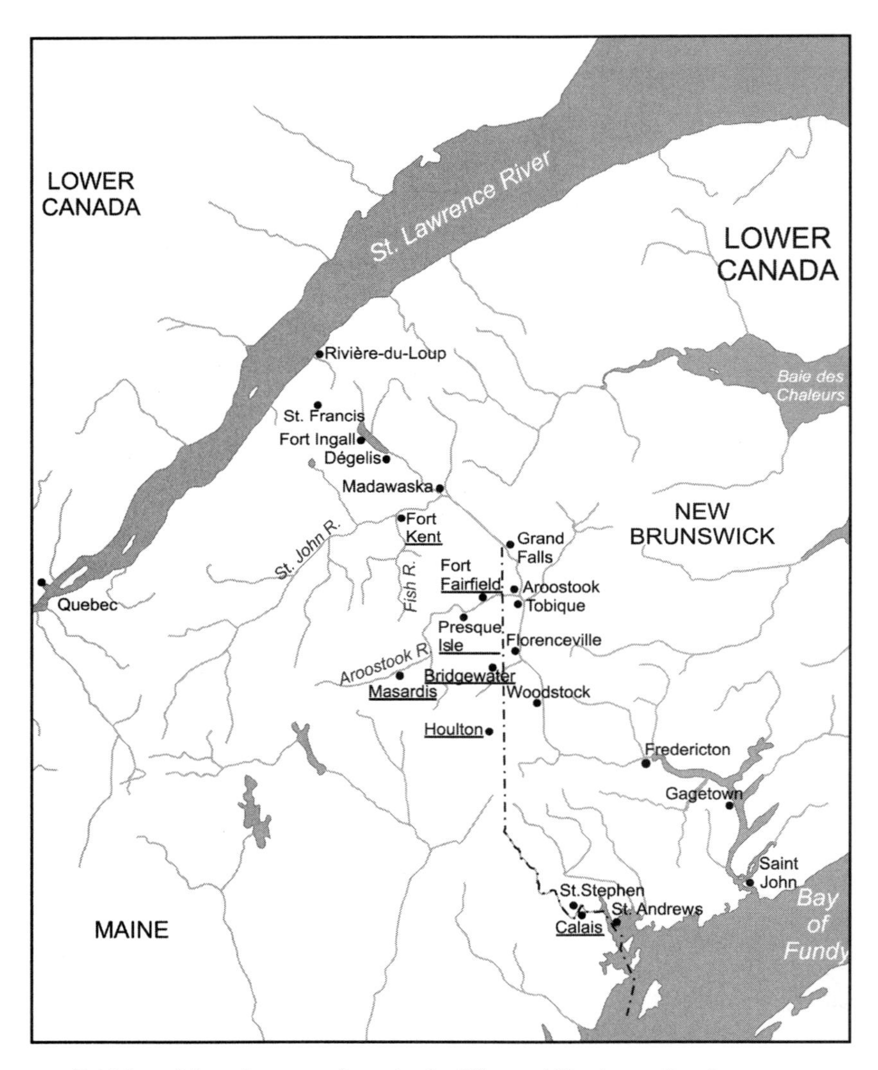

British and Americans garrisons in the Disputed Territory after the Aroostook War. American posts are underlined.

timber from encroachments by establishing a boom and "seizing officer" on the Aroostook River. If the posse remained, then Harvey would use force to prevent "an interference with . . . [the British] . . . possession and jurisdiction" of the Disputed Territory. Fairfield's reply restated Maine's right to continue its actions and warned that Maine would meet any use of force with force. Acting in support of Maine, the United States Congress passed a bill on March 3, 1839, authorizing the president to call out up to 50,000 militiamen. Surrounding states and provinces offered their support should a "collision" occur. As diplomatic notes and protests flew among Fredericton, Augusta, London and Washington and war fever raged on both sides of the border, military forces began to assemble.

Fairfield appointed Colonel Charles Jarvis as the acting Maine land agent in place of McIntire and sent him to the Aroostook with "about 600 good and effective men, making the whole force now about 750." Jarvis set about building a boom across the Aroostook River at Fort Fairfield and improving the roads into the area. On February 19, 1839, Maine ordered a draft of 10,343 men from the state militia to be readied for active service to support the land agent. Over the next month, a total of 2,904 Maine militiamen were called to active duty. The American army garrison of 120 artillerymen at Hancock Barracks in Houlton remained neutral throughout the crisis, although Fairfield had asked President Van Buren for their assistance.

Meanwhile, Harvey took similar actions. In a series of Militia General Orders, the militia battalions of Carleton, York and Charlotte counties and the Saint John City Militia drafted up to one fourth of their strength for service. Of this, it appears that only 950 troops were actually called out. The 3rd Battalion, Carleton County Militia at Madawaska was excluded from the draft, due, no doubt, to concerns about their loyalties, given the destabilizing influence the Maine agents had had in the area. Harvey also had 1,398 British regulars at his disposal.

Each side made troop dispositions intended to check the other's movements, seeking to avoid a collision and yet prevent the other side from gaining an advantage. By February 15, Harvey had begun deploying his forces. He ordered four companies of the 36th Regiment, the garrison regiment, to move north from Fredericton. He stationed two companies at Woodstock to guard against the Americans at Houlton, and sent another company to Tobique and the mouth of the Aroostook to counter the posse at Fort Fairfield and guard the boom across the Aroostook River. A fourth company went to Grand Falls, where it could guard the Madawaska Settlement and the route to Quebec. The local militia reinforced the regulars. The 1st Battalion of the Carleton Militia posted three companies at Woodstock and one at Buttermilk Creek (Florenceville). The 2nd Battalion, Carleton Militia had one company at Tobique and the mouth of the Aroostook and one at Grand Falls. Detachments of the Royal Artillery and the New Brunswick Militia Artillery were stationed with these forces, armed with 12-pounder howitzers and 6-pounder guns; the detachment at Madawaska even had rockets. The Carleton Light Dragoons and the York Light Dragoons formed a line of picquets between Madawaska and Fredericton to carry dispatches. Harvey's strategy was "confined to the protection of the communication between this province and Lower Canada through the valley of the St. JohnRiver and of Her Majesty's subjects of the Madawaska Settlement." However, he did not have enough troops to carry out these objectives, so he called for help. Sir John Colborne sent the 11th Regiment and a detachment of artillery from Quebec to garrison the Madawaska settlement and the portage route to the St. Lawrence, and Sir Colin Campbell sent the 69th Regiment, which had just arrived in Halifax from the West Indies.

Meanwhile, Major-General Isaac Hodsdon's force of 1,069 Maine militiamen had begun arriving in Houlton on March 5. As Hodsdon advanced toward Fort Fairfield, he left garrisons in Bridgewater and Presque Isle, Maine. Between March 13 and 15, his force arrived in Fort Fairfield and began replacing the posse that was released from service. Another force of about five hundred militiamen under Brigadier-General George W. Bachedler moved up the interior route and arrived at Marsardis on March 12. Major-General Ezekiel Foster mustered a force of 369 men in Calais in the first week of March. Meanwhile, the posse had been actively pursuing trespassers along the Aroostook and Fish rivers and along the Rivière-du-Chute; they appear to have acted fairly peacefully. By mid-March, bodies of troops lined each side of the border extending from St. Andrews and Calais to the Madawaska. Fortunately, no collisions or actual fighting occurred between the armies; the only skirmishes took place between the posse and the trespassers. Generally speaking, both sides kept well apart, notwithstanding an apocryphal story of American and British sentries who stood guard along the Presque Isle River, a branch of the St. John, on opposite sides of the same hill. The only casualty is said to have been a farmer in Fort Fairfield who was struck by a ricochet during victory celebrations. Otherwise, the sobriquet of the "bloodless Aroostook War" appears apt.

While soldiers prepared for battle, diplomats worked out a solution in Washington. Neither country wanted a war, but Britain was prepared to defend the Grand Communications Route at any cost. John Forsyth, the American secretary of state, and H.S. Fox, the British envoy, reached a compromise on February 27. The agreement called on Fairfield to withdraw the Maine militia from the Disputed Territory while the posse remained in the Aroostook valley to prevent trespass. In turn, Harvey would not expel the Maine militia by force and would agree to the presence of the posse in the Aroostook valley. Although both national governments endorsed this plan, Maine did not. General Winfield Scott took on the task of convincing Maine to accept the proposal. Fortunately, Scott and Harvey had known and trusted each other since the War of 1812. Having received instructions from London, Harvey quickly agreed to the proposal. Scott, on his part, convinced the Maine politicians to agree as well. A key section of the agreement stated that Maine would not "disturb by arms the said Province [of New Brunswick] in the possession of the Madawaska settlements, or . . . attempt to interrupt the usual communications between the Province and Her Majesty's Upper Provinces." An exchange of notes between Scott, Harvey and Fairfield on March 21, 23 and 25 sealed the agreement. On March 25 and 27, orders were issued discharging the militias of both Maine and New Brunswick. The discharges would be phased over a period of time, and a group of Maine militia would remain in the Disputed Territory until the Maine land agent organized a suitable civil force. The British garrisons

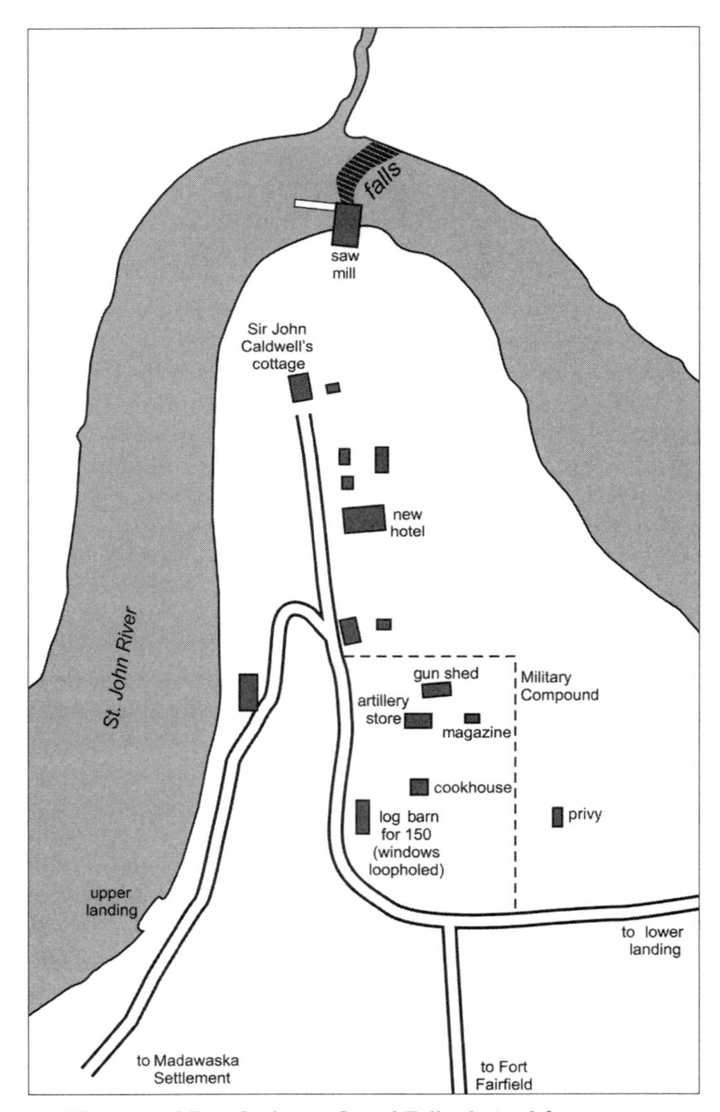

The second Fort Carleton, Grand Falls, derived from a sketch by Major Foster, R.E. (c. 1839). The rebuilt fort was meant to be defended, as indicated by the loopholed windows in the log barn barracks.

stayed at Woodstock and the mouth of the Aroostook until June. The Grand Falls garrison remained in place to guard the upper St. John River. Harvey also assured Scott that he would take measures to protect the timber in the areas still under British jurisdiction, issuing the appropriate orders to MacLauchlin. The final resolution of the border controversy would be worked out in due course.

The 11th Regiment left the Madawaska settlement around the end of March, 1839. At about the same time, Colonel Jarvis sent Captain Alvin Nye, with some twenty-three members of the Maine posse, to the Fish River to stop timber depredations. In the meantime, Jarvis completed the boom across the Aroostook River and, on April 9, began construction of two blockhouses at Fort Fairfield. In early April, Nye arrived at Soldiers' Pond, about four miles upstream from the mouth of the Fish River, and began constructing a boom and a blockhouse. By April 23, he had moved to the mouth of the Fish River (Fort Kent) and started a second blockhouse and boom. These measures placed the posse in a firm position on the west (or southern) bank of the St. John River, about twenty miles upstream from the Madawaska River and at the western limits of the established Madawaska Settlement. This activity, in clear violation of the agreement between Scott and Harvey, set off a round of diplomatic protests. While the Americans agreed that Nye's actions were contrary to the agreement, no one on the American side was prepared to take any action. So, despite the agreement, Maine extended its jurisdiction into the Madawaska settlement, an action that threatened both the communications route and the British inhabitants.

Working with Sir John Colborne in Lower Canada, Harvey took steps to secure the communications route that marked the beginning of the consolidation or garrison period of the route's history. By July 25, 1839, Harvey reported that he and Colborne were improving the route and establishing a series of posts along it to facilitate the movement of troops. In June, 1839, Colborne had established a small garrison of thirteen soldiers at Fort Ingall, at Cabano. In 1839, a barracks for two hundred men was built at Fort Carleton at Grand Falls. Then, in 1840, the way station at Dégelis was expanded to house 164 men and Fort

An aerial view of the strategic location of the the blockhouse at Edmundston, built in 1841, which could control traffic on both the St. John and the Madawaska river. Destroyed by fire in 1855, it was reconstructed in 2000. RGA, GC

Ingall at Cabano was improved. In the fall of 1841, a blockhouse was built at the mouth of the Madawaska River. (In 2000, this blockhouse was reconstructed on the site.)

These activities set off successive rounds of protests by the Americans, who claimed the British were violating the terms of the agreement. Indeed, tension had been building once again, provoked by these measures and other confrontations. On the night of September 8, 1839, a group of New Brunswick lumbermen broke into the militia arms storehouse in Tobique and made an abortive raid on Fort Fairfield. In November, heightened tension led to reinforcement of Fort Ingall by one company of infantry from the 11th Regiment, followed by a second

Towboats on the Madawaska (c. 1840), by Mrs. Arianne Saunders Shore. The outpost at Dégelis, shown in the background, was created in 1837 as a way station for troops en route to Quebec and was expanded in 1840. BAG 1969.23

company in December. Clearly, each side had begun marking out its territory in anticipation of new border discussions.

The next crisis came in the fall of 1840, when Maine held two town meetings at Fort Kent. Captain Stover T. Rines, who was in charge of the posse, threatened Mr. Francis Rice, a British peace officer, who tried to protest this exercise of American jurisdiction within the British part of the Disputed Territory. This action pushed British authorities too far, and Harvey asked Lord Sydenham, the governor-general of Canada, to send troops into the Madawaska Settlement to give "support to the civil authorities of the Queen and protection to Her Majesty's subjects in the Madawaska Settlement." In June, a company of the 56th Regiment from

Lower Canada arrived at Fort Ingall and a small detachment occupied Dégelis, and, in December, two companies of the 56th were dispatched to Madawaska. Fairfield, of course, sent protests to Harvey and United States President Van Buren, but without success. At some point during 1841, a detachment of seven soldiers was posted on the south bank of the St. John at Madawaska in a feeble attempt to reassert British jurisdiction in the area.

The final step taken in the garrison period was the occupation of Fort Fairfield and Fort Kent by one company each from the 1st Artillery Regiment of the United States Army in September, 1841. Maine pushed for this measure because its land agent found maintaining the posse too expensive and wanted to pass on this expense to the United States government. At first, the British protested that the troops' presence would give permanence to the American occupation of the territory, but incidents such as the desertion of seven soldiers from the 56th Regiment at Madawaska in April, 1841, caused them to rethink this position. John Baker and three others were arrested and convicted of aiding the desertions, and Captain Rines, the commander of the posse, was implicated as well. Pending a border settlement, Sir William MacBean George Colebrooke recommended that garrisons of regular troops, British and American, in the Disputed Territory would provide a "guarantee for the strict observance" of the existing agreements that did not exist while the posse formed the garrison.

This decade of tension had convinced both the British and Americans that the border controversy must be settled. Accordingly, a new commission headed by Daniel Webster, the U.S. Secretary of State, and Alex Baring, Lord Ashburton, met in Washington, D.C. throughout the summer of 1842. After long negotiations, they reached a compromise. The agreement, known as the Webster-Ashburton Treaty, was signed in 1842 and ratified in 1843. While the United States received 217,301 fewer hectares than had been awarded by King William I, Maine gained access to the St. John River in order to float timber to market. The important outcome was that the British maintained control of the vital Grand Communications Route; Ashburton had received direction from London that compromise on this point was unacceptable, and he achieved this goal. Unfortunately, in order to secure the agreement, the British relinquished control of the south bank of the St. John River in the Madawaska settlement. Those loyal Acadians living there now found themselves residing in the United States. Subsequently, both governments demilitarized the border. American troops were gone by 1845; the British garrisons in Madawaska and Fort Ingall left in September, 1843, while the garrisons in Woodstock and Grand Falls remained until December, 1847.

After sixty years of controversy, the border between Maine, New Brunswick and Quebec had finally been settled and the road to Canada secured. This agreement proved to be a lasting arrangement, for, while the Oregon Boundary Crisis threatened war in the mid-1840s and the border defenses in Canada were strengthened, the Maine-New Brunswick border remained quiet. Had the American Civil War not occurred, it is likely that the role of the Grand Communications Route as a strategic military road would have faded into history.

British Troops Conveyed Through Canada. Six or eight men to a sleigh, and warmly dressed, the troops were conveyed along the Road to Canada by locally engaged teamsters during the Trent Affair. ILN UNB

Chapter Five

British Strategy Vindicated: 1845 to 1870

"[Arrangements for the troop movement] showed greater perfection of our commissariat and medical departments, and the higher ability of our staff, when compared with the Crimea and other campaigns."

— *Lieutenant Francis Duncan,* Our Garrisons in the West, *London, 1864*

In the years following the signing of the Webster-Ashburton Treaty, the Grand Communications Route continued to be used for mail, commerce, private travel and local troop movements. The physical means of transportation improved. In the 1830s, the road from Fredericton to Saint John was shortened when a shortcut was built from Welsford to Oromocto via Petersville (now Route 7). By 1840, a road was opened as far as Grand Falls and was gradually extended to Rivière-du-Loup. On the water, canoes gave way to towboats and, later, steamers. By 1860, one railway ran from St. Andrews to Canterbury, with plans to extend it to Quebec, and another ran from Saint John to Shediac. Also, a telegraph line followed the Grand Communications Route from Saint John to Quebec. Nova Scotia built a road and rail network that linked Saint John, via steamer and the Annapolis Valley, with the imperial military base at Halifax. Although international crises occurred between Great Britain and the United States, including the Oregon Boundary Crisis in the

Troops leaving England ten days after the news of the Trent Affair reached London. ILN UNB

1840s, none was sufficiently serious to require the military use of the Grand Communications Route. Thus there was little to justify the British strategy that had led to the loss of Aroostook County and the Acadian settlements on the south bank of the St. John River at Madawaska. Critics of the Webster-Ashburton Treaty made much of this shortcoming, and in the period of peace that followed, it was hard to fault their logic. All of these considerations changed during the winter of 1861-1862, however, when British strategy was vindicated.

When the American Civil War broke out in 1861, Britain and its North American colonies officially remained neutral, although they tended to be pro-Confederate. Because a long-standing goal of some northern U.S. politicians was to add Canada to the Union, concern arose that American forces would move north after their anticipated quick victory over the Confederate states. Other than reinforcing Canada with three battalions of infantry and a battery of field artillery during the summer of 1861, the British adopted a "wait and see" policy. This period of cautious tension ended abruptly on November 8, 1861, when Union sailors from the USS *San Jacinto* boarded the British mail steamer *Trent* in the Bermuda passage and forcibly removed two Confederate commissioners who were en route to Britain and France. The Trent Affair enraged the British government and people. The obvious American glee over having twisted the lion's tail further inflamed the crisis. War seemed certain. While the British demanded the release of the commissioners and the dying Prince Albert tried to find a peaceful solution, the War Office in London quickly implemented plans for the immediate reinforcement of British North America. News of the Trent Affair reached London on November 28, the decision to reinforce Canada was taken on December 6, and the first troops sailed the next day.

The Grand Communications Route through New Brunswick proved essential for reinforcing Canada. Troops arrived in the Nova Scotia command directly by sea. However, the majority were allocated to Canada, which presented a serious problem. The St. Lawrence closed in late November, although it was hoped that ships could get as far up as Rivière-du-Loup, the eastern terminus of the Grand Trunk Railway, or Bic, some eighty-seven kilometres east of Rivière-du-Loup. If this was not possible, the ships would divert to Halifax. The troops would then proceed to Saint John and move overland to Rivière-du-Loup, where they would take the train to points west. Although quickly planned, the deployment was well organized, the War Office having learned hard lessons about poor logistical planning from the disastrous winter of 1854-1855 during the Crimean War. Officers with experience in Canada were consulted extensively, as was Florence Nightingale, who gave valuable advice on the sanitation and health of the troops during a winter march. Deputy Commissary General Thomas Wilson, a retired commissariat officer who had planned the march of the 43rd, 85th and 34th Regiments during the winter of 1837-1838, made many suggestions concerning transportation, clothing, accommodation and rations that were incorporated into the final plan.

Hectic preparations took place in the United Kingdom as the British warned troops for duty, chartered ships and readied supplies of weapons,

The side-paddle steam transport *Adriatic* in the ice at Sydney, Cape Breton. ILN UNB

ammunition, camp stores and uniforms for both the British regulars and the militia in British North America. The winter voyage across the stormy North Atlantic was fraught with danger, and troopships and their escorts lost touch during bad weather. Most of the chartered ships were side-paddle steamers, which limited their ability to navigate the ice in the approaches to the St. Lawrence River and exposed their engines to storm damage. Only one ship, the *Persia*, actually made it up the St. Lawrence as far as Bic. As the men of the 1st Battalion of the 16th Regiment were disembarking, ice rushed downriver, and the ship quickly had to withdraw, leaving a company of infantry still on board and a portion of the crew stranded on shore; the soldiers helped sail the ship to Halifax. The ship carrying most of the 96th Regiment was damaged and had to withdraw to England after two attempts; and the Admiralty commended the regiment for its efforts in helping to save the ship.

By the end of December, 1861, the Halifax harbour was full of troopships. Normally, winter travellers en route to Canada now took the railway from Portland, Maine, to Montreal, but clearly this was not an option for British troops. They would have to sail to Saint John, and from there two routes were considered. The first was to advance along the Grand Communications Route to Rivière-du-Loup. An alternate course, called the Matapedia Road, followed the rail link from Saint John to Shediac and then the road to Campbellton and overland to Métis on the St. Lawrence. After investigation, this route did not seem practical. So, while British forces assembled in Halifax, the military staff busily arranged for transportation, lodgings and food along the Grand Communication Route.

In addition to transportation, the movement plan also considered enemy action, weather and desertion. Fortunately, the enemy threat receded when, on January 9, 1862, the Union released the Confederate commissioners and the urgency of the crisis subsided. The troops who were already en route continued their journey, but departures from Britain were halted. The Union graciously offered the British force use of the Portland-Montreal railway link; however, the British declined.

An advanced headquarters was established at Saint John to control the movement of the troops. The 1st Battalion Military Train managed the transportation. Assistant Commissary General Edmund M'Mahon arranged contracts for transportation at Fredericton, awarding them to three contractors who divided the route into three stages. Mr. Orr took over the first stage from Saint John to Fredericton, while Major James Rice Tupper looked after the next one, from Fredericton to Little Falls (Edmundston). Mr. A.A. Glazier received the contract for the last stage, from Little Falls to Rivière-du-Loup. These contractors provided roughly constructed two-horse sleighs capable of holding eight men; the Guards, being larger men, travelled six to a sleigh. Each sleigh carried a small repair kit consisting of a saw, hammer, nails, clasp knife and cord, plus an allocation of snow shovels and snowshoes. Many of the drivers and horses normally worked in the lumber trade or farming and so were

Armstrong guns packed on sleighs in the Saint John ordnance yard, ready to be hauled over the Grand Communications Route. ILN UNB

familiar with winter conditions. Similar sleighs provided carriage for the eighteen Armstrong guns of the three artillery batteries.

The troops divided into packets of approximately 160 men for the trip. A typical packet consisted of a sleigh carrying half of the officers, baggage sleighs with an escort, and sleighs with the main body of troops; the last sleigh carried the remaining group of officers. Before leaving Britain, each soldier received cold-weather clothing: fur caps with ear lappets, woollen comforters, chamois waistcoats, a flannel shirt and drawers, warm gloves, a pair of long boots and thick woollen stockings. In addition, the men obtained moccasins at Saint John, and the contractors provided straw and buffalo robes for use in the sleighs. For further warmth, the men ate three hot meals a day, and they were

Arrival of a detachment of the 63rd Regiment at the temporary barracks at Petersville, the first stop on the Great Communications Route and now part of Canadian Forces Base Gagetown. The officers' quarters are shown at the left, the cookhouse is in the centre, and the soldiers' barracks are at the right. ILN UNB

encouraged to run alongside the sleighs in shifts to maintain their circulation. Medical officers travelled with most groups and others waited at each of the halts.

The route had improved considerably since the previous movements in 1837-1838. It now followed an established road, although the section through New Brunswick was not as good as the portion in Canada. Snowplows and rollers kept it open during the bad weather. Barring weather delays, the journey by sleigh was expected to take ten days. The timetable included nine overnight stops, manned by support troops. Where possible, the troops were billeted in existing buildings, such as houses, hotels, warehouses or barns, as well as in the permanent barracks in Saint John and Fredericton and the abandoned post at Fort Ingall, which had

ROUTE OF THE OVERLAND MARCH

Day	Distance (Miles)	Location	Remarks
0	0 (0)	Saint John	Controlling Headquarters. Major General Rumley commanding
1	30 (30)	Petersville	Assistant Surgeon and two men of the Commissariat Staff Corps (CSC)
2	30 (60)	Fredericton	Officer and detachment of the Military Train (MT), medical officers and detachment of the Army Hospital Corps (AHC), Assistant Commissary General and detachment of the CSC
3	29 (89)	Dumfries	Two men of the CSC
4	32 (121)	Woodstock	Detachments of the CSC and MT and two men of the AHC
5	23 (144)	Florenceville	Assistant Surgeon, an officer and two men of the MT, and two men of the CSC
6	26 (170)	Tobique (Andover)	Assistant Surgeon, an officer and two men of the MT, and two men of the CSC
7	24 (194)	Grand Falls	A Staff Surgeon, an officer and two men of the MT, an Assistant Commissary General and two men of the CSC
8	6 (230)	Little Falls (Edmundston)	Assistant Surgeon, an officer and two men of the MT, and two men of the CSC
	Mid-day stop	Degele (Dégelis)	
9	37 (267)	Fort Ingall (Cabano)	A Staff Surgeon and two men of the AHC, an officer and two men of the MT, an Assistant Commissary General and three men of the CSC Rations for 200 men for 30 days stocked here
	Mid-day stop	St. Francis	Two men of the CSC Rations for 200 men for 5 days stocked here
10	42 (309)	Rivière-du-Loup	An Assistant Quartermaster General, a Staff Surgeon and four men of the AHC, an Assistant Commissary General and five men of the CSC. Transfer to Grand Trunk Railway.

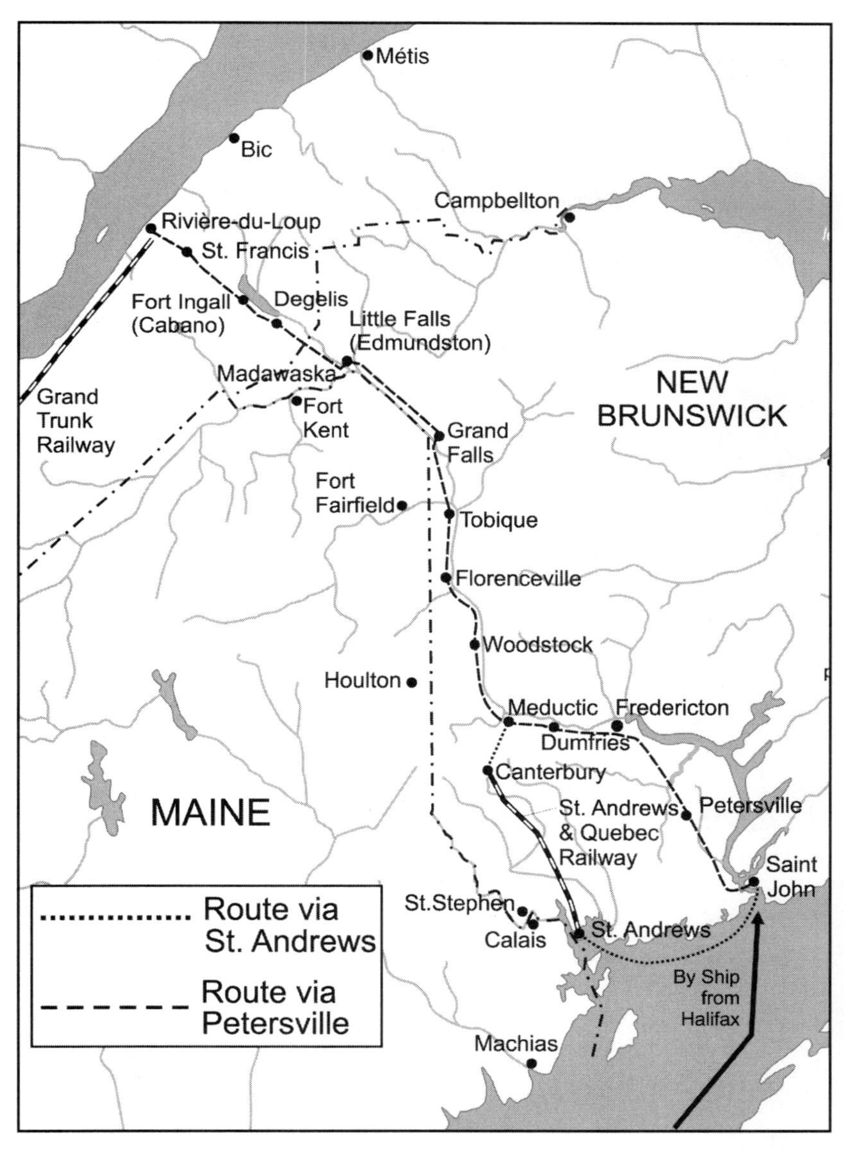

Movement of troops and supplies over the Grand Communications Route during the Trent Affair, January to March, 1862.

St. Andrews harbour, the terminus of the St. Andrews and Quebec Railway. ILN UNB

been refurbished. It was only at Petersville and St. Francis that temporary long, low log buildings called cabanos were built for shelter. The officers stayed in nearby hotels or private homes. They purchased food locally, although the Commissariat Department set up bakeries at Grand Falls and Fort Ingall. Remembering the experience of the company of the 104th Regiment that was storm delayed in 1813 around Lake Temiscouata and ran out of food, planners established reserve stores of food. All told, the force had a much easier trip than the 104th, who had to march pulling toboggans.

Troops began leaving Halifax for Saint John on January 1, 1862. The first to depart was the 62nd Regiment, which was headquartered in Halifax with detachments in Fredericton, Saint John and St. John's, Newfoundland. Its role was to secure the route from any possible American interference, especially by units of the Union Army at Houlton, Maine, and garrison the stopping places if required. If necessary, the 1st Battalion of the Rifle Brigade would have supported them, but fortunately the Americans did not interfere. The St. Andrews and Quebec Railway ran as far as Canterbury, and this appeared to be such a good way to move troops that the 62nd, along with an ad hoc battery of field artillery and a third of the 1st Battalion of the Military Train, sailed from Saint John to St. Andrews and embarked by train. From Canterbury, they proceeded by sleigh to Woodstock and then onwards up the river. However, the railway was unequal to the task due to the cold and the large quantity of snow on the tracks, and therefore by mid-January the St. Andrews route was abandoned. The army fell back on the traditional path of the Grand Communications Route, and the rest of the troops left directly from Saint John, the forward headquarters.

Dr. H. Chalmers Miles, a surgeon with the artillery batteries, described his departure.

> Conceive a dark, dreary morn, with just a streak of daylight in a dull leaden sky, with the snow descending in heavy wet flakes; before you a long row (thirty or more) of rough-looking unshapely trucks, with long, wide, wooden runners on either side; put upon these trucks four deal seats; place two men in each seat, with their blanket-sacks drawn over their feet and legs, the extra blanket over the head and shoulders, and the remainder of the body concealed by buffalo robes; realize to yourself that they are generally shapeless objects, as they sit huddled together, a long cloud redolent of bad tobacco steaming above them; notice that to each rude vehicle, with its queer freight, is fastened a pair of large, raw-boned, ill-conditioned horses, and that the wild-looking object muffled in furs is the "teamster"; imagine that every man in each of those thirty trucks has put on every possible article of warm clothing in his possession, — and you will have before you a very tolerable idea of the sort of scene represented as our party started in sleighs from St. John.

For the first time, each group of troops reported in every evening by telegraph. Their travel was controlled on the basis of these reports, and subsequent departures were delayed if preceding groups were held up by storms, as occasionally happened. Staff officers, constantly moving up and down the route in express sleighs, exercised additional control. By March 13, 1862, the last group of troops had cleared Rivière-du-Loup. The troops were received with great warmth and kindness all along their route, which greatly eased their passage. In all, 274 officers and 6,544 men passed along the route, plus eighteen guns and the equipment of the three field batteries of artillery. An unrecorded quantity of military stores was also transported. This was the largest troop movement ever mounted along the Grand Communications Route.

The following regiments and other units made the overland march to Canada during the winter of 1861-1862:

> *Infantry*: 1st Battalion Grenadier Guards, 2nd Battalion Scots Fusilier Guards, one company 1st Battalion 16th (Bedfordshire) Regiment, 62nd (Wiltshire) Regiment, 63rd (West Suffolk) Regiment, 1st Battalion The Rifle Brigade
>
> *Artillery*: E, F, and G Batteries 4th Brigade Field Artillery, Numbers 5 and 6 Batteries 7th Brigade Garrison Artillery, Numbers 1, 4, 5 and 6 Batteries 10th Brigade Garrison Artillery
>
> *Engineers*: Number 15 and 18 Companies Royal Engineers
>
> *Support Corps and Others*: 56 cavalry instructors for cavalry and volunteers, 1st and 3rd Battalions Military Train, detachments of medical officers and men of the Army Hospital Corps, detachments of Commissariat Department, officers and men of the Commissariat Staff Corps

In addition, the reinforcement of New Brunswick consisted of the 1st Battalion of the 17th Regiment, four companies of the 96th Regiment, A

Battery of the 8th Brigade Field Artillery, H Battery of the 4th Brigade Field Artillery and 4 Company Royal Engineers.

Because of the excellent medical arrangements, few casualties occurred during the move. Not more than seventy men entered the hospitals en route; only two died as a result of disease, and another two died from excessive drinking. Of the eleven cases of frostbite, only one was serious and that was because the victim had also been drinking to excess. Although temperatures fell as low as -32°C, the winter was considered mild because there was little wind. The only serious delay was caused by a blizzard on January 21-23. Desertion was minimal. American recruiters, called crimps, were very active along the Maine-New Brunswick border, looking for trained soldiers for the Union Army. They offered British soldiers tempting bounties and promotion if they would desert and enlist in the Union Army. The town of Tobique (present day Andover) became a particular hotbed of this activity. The lieutenant-governor of New Brunswick called out the militia to help guard against the crimps as well as assist with the movement of the troops. The officers travelling in each of the packets were specially charged to be on their guard. In total, there were only nine desertions, three of these at Tobique

By the time the troops arrived in Canada, the threat of invasion by the Union had dissipated, and a reduction in the force level followed during the summer of 1862. However, Anglo-American relations remained tense until outstanding differences were finally settled with the signing of the Treaty of Washington in 1871.

The Grand Communications Route had once more proved its strategic importance during the Trent Affair. However, even with restored good relations with the Americans, caution prevailed. When the Intercolonial Railway was built in the 1870s, its route between Quebec and Halifax took the roundabout way through the Gaspé Peninsula and along the east coast of New Brunswick, as far away as possible from any potential American threat. Ironically, some of the money used to build the railway had originally been allocated to Canada to fortify Saint John in order to defend the Grand Communications Route. With the rail link between Halifax and central Canada established, and the improvement of winter navigation in the St. Lawrence, the strategic importance of the Grand Communications Route declined and was soon forgotten.

Epilogue

The story of the Grand Communications Route lies at the heart of New Brunswick's history. Great rivers shaped patterns of human occupation from the earliest arrival of the First Nations until the late nineteenth century. European settlement in particular was based on, or influenced by, military considerations linked to communications and borders. The international border between Maine and New Brunswick resulted from a compromise that ensured British control over the Grand Communications Route. The strategic importance of this route as the key overland line of communication between Canada and the outside world was confirmed on many occasions throughout its long history.

The military significance of this route faded as technology changed and friendly relations between the United States and Canada developed. However, part of the Grand Communications Route remains a crucial national artery for commercial and private travel. A comparison of the route of the modern Trans-Canada Highway and the portage routes of the First Nations shows similarities. The Trans-Canada still follows the Grand Portage from the St. Lawrence to Lake Temiscouata and then the Madawaska and St. John rivers to Oromocto. From there, the Trans-Canada Highway continues eastward to Fort Beauséjour and Nova Scotia, paralleling the Lake Washademoak-Canaan River-Petitcodiac River portage route. Highway 7, running south to Saint John and the sea connection to Nova Scotia, follows the path of the Grand Communications Route. The Woodstock-to-Houlton road leads to Interstate 95 and southern Maine as did the great Eel River portage route.

The Grand Communications Route is used frequently by the convoys of military vehicles travelling between bases at Petawawa, Ontario, and Valcartier, Quebec, and the large training area at Camp Gagetown,

The Officers' Quarters in the Military Compound, Fredericton, is a major tourist attraction. The York-Sunbury Historical Society Museum is located in this building. COURTESY FREDERICTON TOURISM

which is a major training and support base. Saint John remains an important garrison town and is the home of several reserve units, and military garrisons still exist along the route at Edmundston, Grand Falls, Woodstock and Fredericton. Remnants of the military establishments and fortifications have survived at St. Andrews, Saint John, Fort Beauséjour and along the St. John River at Fredericton and Grand Falls. The blockhouses at Oromocto and Edmundston have been reconstructed, as has Fort Ingall at Cabano, Quebec. Historical plaques mark many other sites.

Residents and visitors to New Brunswick continue to use the Grand Communications Route every day. Hopefully, this book will help them appreciate the vital role New Brunswick played in defending the road to Canada and how the defence of empire shaped the history of the province.

Fort Ingall. Located at the start of the Grand Portage in Cabano, Fort Ingall has been partially reconstructed as an informative tourist attraction. COURTESY LA SOCIÉTÉ D'HISTOIRE ET D'ARCHÉOLOGIE DE TIMISCOUATA

Grand Falls Barracks. This building, now part of the Senechale Furniture Company, is believed to be the barracks built in 1839. GC

Historic Sites to Visit along the Grand Communications Route

Saint John. Many of the forts, batteries, buildings and the Martello Tower still exist and can be visited. A full description of them is provided in *Saint John Fortifications, 1630-1956,* Volume 1 in the New Brunswick Military Heritage Series.

Loyalist Settlements. More information about these settlements is contained in *Hope Restored: The American Revolution and the Founding of New Brunswick*, Volume 2 in the New Brunswick Military Heritage Series.

Nerepis. A plaque on Highway 102 at Woodman's Point, at the junction of the St. John and the Nerepis rivers, marks the site of Fort Boishébert.

Worden's Point. The remains of the old French battery and the War of 1812 British battery can still be seen at Eagle's Nest on Highway 124, just across the St. John River from the Evandale ferry.

Oromocto. The Fort Hughes blockhouse has been rebuilt on the bank of the Oromocto River at Sir Douglas Hazen Park, beside the Oromocto Shopping Centre.

Fredericton. The historic Military Compound preserves many of the original buildings, including the Soldiers' Barracks, the Guard House and the Militia Arms Storehouse. The Officers' Quarters houses the York-Sunbury Historical Society Museum.

Prince William. King's Landing Historic Settlement has preserved a number of historical homes and other buildings that would have been lost when the headpond of the Mactaquac Dam was flooded in the late

1960s. Some of the buildings may have been used as way stations by the postal couriers or troops passing along the Grand Communications Route. Each summer, one weekend is dedicated to a recreation of the border tensions of the 1830s.

Presqu'Ile. A historic plaque marks the site of the post. It is located on Highway 103 at the junction of the Presqu'Ile and St. John rivers near Simonds.

Grand Falls. The warehouse of the Senechale Furniture Company on Portage Road is believed to be the barracks that was built in 1839.

Edmundston. The blockhouse, rebuilt in 2000, stands high above the junction of the St. John and Madawaska rivers.

Cabano, Quebec. Fort Ingall, partially rebuilt, is a museum operated by the Société d'histoire et d'archéologie de Timiscouata.

Acknowledgements

Any project like *The Road to Canada* is the result of the combined efforts of a number of people. I first learned of the Trent Affair while researching the military service of Driver Frederick McKenzie, Land Transport Corps, and his subsequent service with the 3rd Battalion Military Train in Canada. The story of the Grand Communications Route developed from there. Both David Wilson and Bill Parenteau at the University of New Brunswick indulged my interest in this subject and encouraged my research into it during my post-graduate courses.

Many people have assisted me. Richard Jamer, Mary Doughty, Terry McCormack and Patrick McCooey have guided me to the scenes of the Aroostook War and the sites of the Upper Posts. More recently, once the New Brunswick Military Heritage project commissioned this book, Robert Dallison and Ernest Clarke kindly shared with me the results of their research. Doug Knight skilfully tracked down records at the National Archives of Canada for me. The members of the Crimean War and Medals Internet discussion groups have consistently been excellent sources of information. Equally helpful have been the staffs at the Harriet Irving Library at the University of New Brunswick, the Provincial Archives of New Brunswick, Nova Scotia Archives and Records Management, and the New Brunswick Museum, as have members of the York-Sunbury Historical Society, the Carleton County Historical Society and the Grand Falls Historical Society. The unfailing cooperation of all of these individuals and organizations made my task that much easier.

Marc Milner and Brent Wilson of the New Brunswick Military Heritage Project have patiently edited my drafts and provided excellent suggestions for revisions. Mike Bechthold has done an amazing job of turning my rough sketches into outstanding maps. Laurel Boone and the staff at Goose Lane Editions have done a wonderful job in the final editing and production of this book. In addition to being my sounding board, my wife Carolyn has continually given me her full support during this project. I thank you all.

This book provides only an overview of the history of the Grand Communications Route and the Road to Canada. Much more research could be done, and I hope that *The Road to Canada* will encourage further work in this area. The responsibility for any errors or omissions remains mine alone.

Illustration Credits

Photos and other illustrative material on pages 50 and 69 appear courtesy of the National Archives of Canada (NAC); on page 30 courtesy of the National Gallery of Canada (NGC); on page 79, courtesy of the Beaverbrook Art Gallery (BAG); on page 60 courtesy of the New Brunswick Museum (NBM); on page 40, courtesy of Heritage Resources, Saint John (HRSJ); and on pages 78 and 79 courtesy of Gary Campbell (GC). Engravings on pages 10, 86, 88, 89, 92, and 94 from the *Illustrated London News* appear courtesy of the University of New Brunswick Archives and Special Collections, Harriet Irving Library (ILN UNB). The aerial image on page 78 apears courtesy of Ron Garnett, Airscapes.ca (RGA). The maps on pages 32, 38, 56, 72 and 91 are by Mike Bechthold, NBMHP's cartographer. All illustrative material is reproduced by permission.

Selected Bibliography

Bourne, Kenneth. *Britain and the Balance of Power in North America 1815-1908*. Berkeley and Los Angeles: University of California Press, 1967.

Burrage, Henry S. *Maine in the Northeastern Boundary Controversy*. Portland, ME: Marks Printing Houses, 1919.

Carroll, Francis M. *A Good and Wise Measure: The Search for the Canadian-American Boundary, 1783-1842*. Toronto: University of Toronto Press, 2001.

Clarke, E.A. *The Siege of Fort Cumberland: An Episode in the American Revolution*. Montreal and Kingston: McGill-Queen's University Press, 1995.

Classen, H. George. *Thrust and Counterthrust: The Genesis of the Canada-United States Boundary*. Don Mills, ON: Longmans, 1965.

Dallison, Robert. L. *Hope Restored: The American Revolution and the Founding of New Brunswick*. Vol. 2, New Brunswick Military Heritage Series. Fredericton: Goose Lane Editions and the New Brunswick Military Heritage Project, 2003.

Facey-Crowther, David. *The New Brunswick Militia 1787-1867*. Fredericton: New Ireland Press, 1990.

Ganong, W.F. *Historic Sites in the Province of New Brunswick*. 1899 St. Stephen, NB: Print'n Press, 1983.

Graves, Donald, ed., *Merry Hearts Make Light Days: The War of 1812 Journal of Lieutenant John Le Couteur, 104th Foot*. Ottawa: Carleton University Press, 1993.

Hannay, James. *A History of New Brunswick*. Saint John: John A. Bowes, 1909.

Hitsman, J. McKay, ed., "A Medical Officer's Winter Journey in Canada," *Canadian Army Journal* (October 1958), pp. 46-64.

Jones, Howard. *To the Webster-Ashburton Treaty, A Study in Anglo-American Relations 1783-1843*. Chapel Hill, NC: University of North Carolina Press, 1977.

Judd, Richard W., Edwin A. Churchill and Joel W. Eastman, ed. *Maine The Pine Tree State from Prehistory to the Present*. Orono: University of Maine Press, 1995.

Marquis, Greg. *In Armageddon's Shadow: The Civil War and Canada's Maritime Provinces*. Montreal and Kingston: McGill-Queen's University Press, 2000.

McNutt, W.S. *New Brunswick, A History: 1784-1867*. 1963 Toronto: Macmillan, 1984.

Morrison, James H. "Wave to Whisper: British Military Communications in Halifax and the Empire, 1780-1880," Manuscript Report Number 338, Parks Canada, 1979.

Raymond, W.O. *The River St. John*, 2nd ed. Sackville, NB: Tribune Press, 1995.

Sarty, Roger and Doug Knight. *Saint John Fortifications, 1630-1956*. Vol. 1, New Brunswick Military Heritage Series. Fredericton: Goose Lane Editions and the New Brunswick Military Heritage Project, 2003.

Scott, Geraldine Tidd. *Ties of Common Blood: A History of Maine's Northeast Boundary Dispute with Great Britain, 1783-1842*. Bowie, MD: Heritage Books, 1992.

Squires, W. Austin. *The 104th Regiment of Foot (The New Brunswick Regiment) 1803-1817*. Fredericton: Brunswick Press, 1962.

Stacey, C.P. *Canada and the British Army: 1846-1871*. Toronto: University of Toronto Press, 1963.

Trueman, Stuart. *The Ordeal of John Gyles: Being an Account of his Odd Adventures, Strange Deliverances, etc. as a Slave of the Maliseets*. Toronto: McClelland and Stewart, 1966.

Webster, John Clarence. *Acadia at the End of the Seventeenth Century*. Saint John: New Brunswick Museum, 1934; rpt 1979.

Webster, John Clarence. *A Historical Guide to New Brunswick*, rev ed. Fredericton: New Brunswick Government Bureau of Travel and Information, 1947.

Young, Richard T. *Canadian Historic Sites: Occasional Papers in Archaeology and History, No. 23 - Blockhouses in Canada, 1749-1841*. Ottawa: Parks Canada, 1980.

Index

C

D

E

F

G

H

I

J

K

L

M

N

O

P

T

U

V

W

Y

The New Brunswick Military Heritage Project

The New Brunswick Military Heritage Project, a non-profit organization devoted to public awareness of the remarkable military heritage of the province, is an initiative of the Military and Strategic Studies Program of the University of New Brunswick. The organization consists of musuem professionals, teachers, university professors, graduate students, active and retired members of the Canadian Forces, and other historians. We welcome public involvement. People who have ideas for books or information for our database can contact us through our Web site: www.unb.ca/nbmhp.

One of the main activities of the New Brunswick Military Heritage Project is the publication of the New Brunswick Military Heritage Series with Goose Lane Editions. This series of books is under the direction of Marc Milner, Director of the Military and Strategic Studies Program, and J. Brent Wilson, Research Director of the Centre for Conflict Studies, both at the University of New Brunswick. Publication of the series is supported by a grant from the Canadian War Musuem.

The New Brunswick Military Heritage Series

Volume 1

Saint John Fortifications, 1630 - 1956, Roger Sarty and Doug Knight

Volume 2

Hope Restored: The American Revolution and the Founding of New Brunswick, Robert L. Dallison

Volume 3

The Siege of Fort Beauséjour, 1755, Chris M. Hand

Volume 4

Riding into War: The Memoir of a Horse Transport Driver, 1916-1919, James Robert Johnston

Volume 5

The Road to Canada: The Grand Communications Route from Saint John to Quebec, W.E. (Gary) Campbell

About the Author

Major W.E. (Gary) Campbell is an army officer with over forty-one years of service in the Canadian Army (Militia), Canadian Army (Regular) and the Canadian Forces. As a transportation officer in the Logistics Branch, he has served in a variety of line and staff positions in navy, army, air force and headquarters units across Canada and in the United States and the United Kingdom. He is presently posted to the Headquarters, Combat Training Centre, Camp Gagetown, New Brunswick, and he is a PhD candidate in the history department at the University of New Brunswick.

An active member of the Orders and Medals Research Society, Gary Campbell has served on the boards of the Military Collectors Club of Canada and the York-Sunbury Historical Society. His articles on the history of military logistics have appeared in journals including *The Royal Logistics Corps Review* and *The Army Doctrine and Training Bulletin* and in *A Handbook on the Canadian Forces Logistics Branch*.